O F G KILGOUR

BSc MIBiol AIFST

An Introduction to the BIOLOGICAL ASPECTS of Nursing Science

WILLIAM HEINEMANN MEDICAL BOOKS LTD

23 BEDFORD SQUARE, LONDON WC1B 3HT

First published 1978

ISBN 0 433 18473 6

Text set in 12 pt Photon Imprint, printed by photolithography, and bound in Great Britain at The Pitman Press, Bath

By the same author

AN INTRODUCTION TO THE CHEMICAL ASPECTS OF NURSING SCIENCE

AN INTRODUCTION TO THE PHYSICAL ASPECTS OF NURSING SCIENCE

AN INTRODUCTION TO SCIENCE FOR CATERING AND HOMECRAFT STUDENTS

AN INTRODUCTION TO SCIENCE AND HYGIENE FOR HAIRDRESSERS (with Marguerite McGarry)

MULTIPLE CHOICE QUESTIONS IN BIOLOGY AND HUMAN BIOLOGY

MULTIPLE CHOICE QUESTIONS IN FOOD AND NUTRITION

MULTIPLE CHOICE QUESTIONS IN HAIRDRESSING AND BEAUTY THERAPY SCIENCE

MULTIPLE CHOICE QUESTIONS IN THE HOUSE AND ITS SERVICES

SHOPPING SCIENCE

AT HOME WITH SCIENCE

EXPERIMENTAL SCIENCE FOR CATERING AND HOMECRAFT STUDENTS (with Aileen L'Amie)

MULTIPLE CHOICE QUESTIONS IN BIOLOGY AND HUMAN BIOLOGY is a useful companion volume for this text

Contents

Chemical Names

IUPAC (the International Union of Pure and Applied Chemistry) recommends chemical names in place of common names. The main chemical names encountered in this book are:

Common current name	Recommended IUPAC name
acetate	ethanoate
acetic acid	ethanoic acid
acetone	propanone
alcohol ethyl	ethanol
alcohol methyl	methanol
bicarbonate	hydrogencarbonate
chloral	trichloroethanal
chloral hydrate	2,2,2-trichloroethanediol
chloroform	trichloromethane
citric acid	2-hydroxypropane 1,2,3-tricarboxylic acid
ether	ethoxyethane
ethyl alcohol	ethanol
ethylene	ethene
fatty acids	alkanoic acids
formaldehyde	methanal
formic acid	methanoic acid
glycerine or glycerol	propane 1,2,3-triol
lactic acid	2-hydroxypropanoic acid
methyl alcohol	methanol
pyruvic acid	2-oxopropanoic acid
urea	carbamide

The IUPAC names define the structure of a chemical compound. The common name gives no indication of chemical structure, these names are still widely used, and are less clumsy than IUPAC names if they are continually repeated.

Preface

I welcome the publication of this book which is designed to introduce nurses in training to the background of science that has particular application to their work. Rapid development in medical technology has made an impact upon patterns of nursing care and it is so often assumed that nurses have a sound knowledge of science, but this is not always so. This book sets out to provide the essential background of pathology, nutrition, microbiology, anatomy and physiology, which will be of great help, not only to nurses in training, but also for the pre-nursing student who has facilities for laboratory work. It is a very readable book and has the advantage of also meeting the basic requirements of the tutors diploma.

It is my hope that this book will help to gain recognition of the science aspect of nursing so that this will eventually become a school subject for study to GCE 'O' Level, alongside those existing Applied Sciences such as Engineering, Building, Domestic and Agricultural Science, which are already GCE examination subjects.

Fred Kilgour, until recently Senior Lecturer in Science at Llandrillo Technical College, Colwyn Bay, has always maintained a keen interest in nursing—his wife Barbara is a trained and practising nurse—and consequently the book is presented with a sensitive appreciation of the needs of nurses in training. The book is very well illustrated, much of the artwork being of Mr Kilgour's own design, and I know that this work will fill a gap in nursing literature which has long been recognised by nurse tutors.

S. G. BADLAND, SRN, RMN,
RNMS, MRIPHH
District Nursing Officer
Clwyd Health Authority, Wales
February 1978

In
memory of Cassie, and to
Gertie and Maud—Nursing Sisters

1 Living Cells

Biology or life science is the study of living things or *organisms*.

CHEMICAL COMPOUNDS AND LIFE

The earth's crust provides the different chemical elements necessary to build up the cells and body of living organisms, for example the percentage chemical composition of a human body is summarised as follows:

- 95% carbon, hydrogen, oxygen and nitrogen
- 4% calcium, phosphorus, chlorine, potassium, sodium and sulphur
- 1% magnesium, iron, cobalt, copper, manganese

Chemical elements combine together forming compounds essential for life activities, the important chemical compounds being *carbohydrates*, *proteins* and *lipids*.

SIGNS OF LIFE

Living things distinguish themselves from non-living or *inanimate* rocks and minerals, in the following most important characteristics:

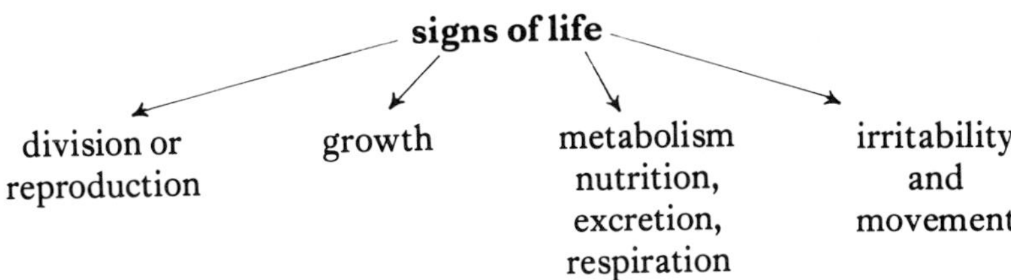

The following summarises the main differences between the living and non-living things:

Living organisms	**Non-living matter**
Plants and animals	Rocks and minerals
1. *Growth* An increase in size of the organism occurs	No growth seen except in crystals

(continued overleaf)

Living organisms Plants and animals	**Non-living matter** Rocks and minerals
2. *Reproduction* Continued maintenance of the plant or animal species or kind	Incapable of reproduction
3. *Metabolism* Chemical changes releasing energy from food, production of waste. Mainly nutrition, excretion, respiration	Chemical changes through chemical reactions under certain laboratory and physical conditions
4. *Irritability* Feeling and response to stimulus of touch, sound, taste, smell, heat, light, electricity or gravity	No response made to an external stimulus
5. *Movement* Ability of certain living things to *contract* their bodies which results in movement of part or all the body	Seen only in machines and motors

CELLS

All living organisms possess the common unit of construction namely the *cell* surrounded by its cell membrane, giving them a *cellular* structure in contrast to the *non-cellular* structure of non-living matter.

Living organisms are composed of either one cell and are *unicellular*, or many cells being called *multicellular*. The study of cells is called *cytology*, whilst *cytopathology* is the diagnosis of disease by examination of cells.

ANIMAL CELL STRUCTURE

The human body is multicellular, a few of these cells can be obtained from the lining of the mouth. This removal of cells from a living body is called a *biopsy* in contrast to the removal of cells from a dead body which is called *necropsy*.

An animal cell, such as the human mouth lining cells, is composed of *protoplasm* is surrounded by the cell wall or *plasma membrane*, made of *lipoprotein* substance. Within the cell is the *cytoplasm* or cell substance and the *nucleus* controlling the life of the cell.

Fluid, mainly water and soluble foods and salts are inside the cell membrane and is called the *intracellular* fluid. Outside the cell membrane is the *extracellular* fluid bathing the cell in a fluid resembling seawater in chemical composition. The animal cell membrane is *selective* in allowing certain substances to pass into the cell cytoplasm, for example it will allow the passage of potassium into the cell but will prevent the entry of sodium.

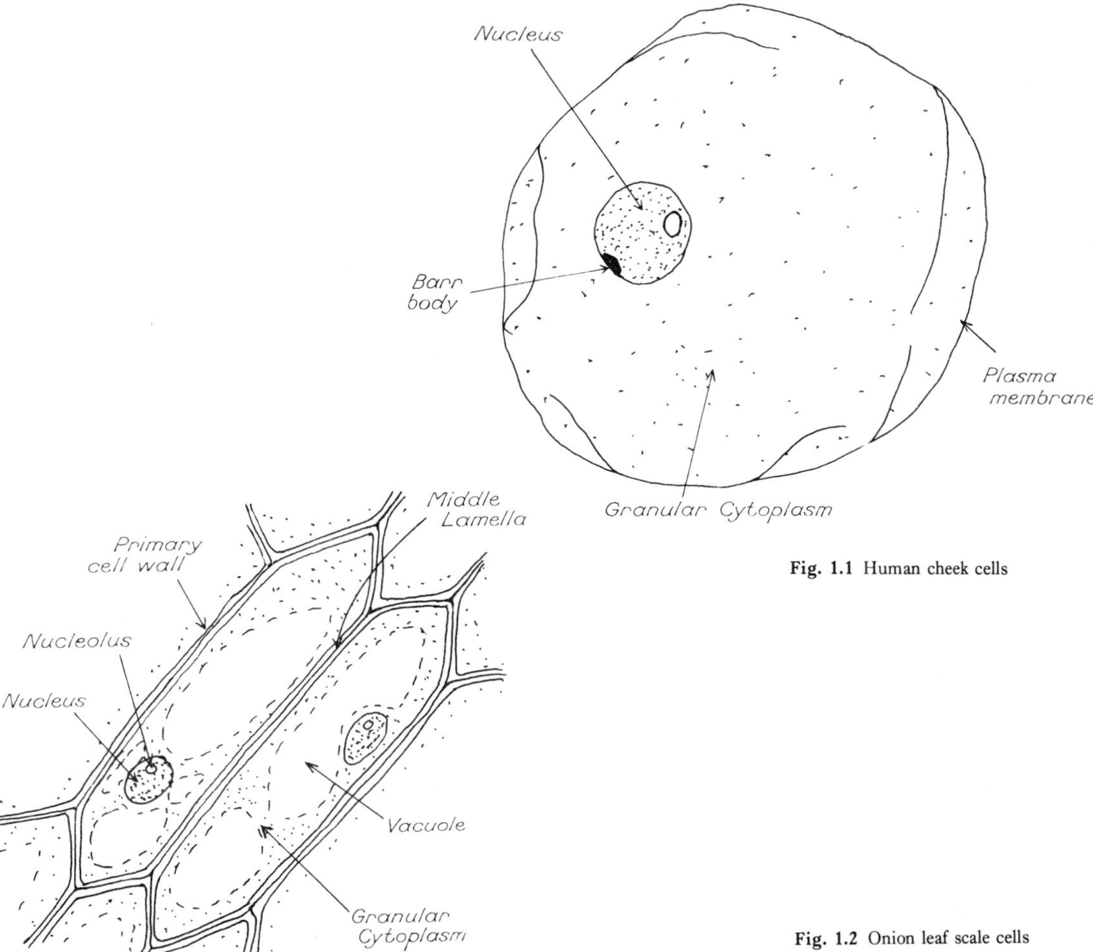

Fig. 1.1 Human cheek cells

Fig. 1.2 Onion leaf scale cells

PLANT CELL STRUCTURE

The onion plant is multicellular and a few cells can easily be examined.

Plant cells have a rigid wall of *cellulose*, between the individual plant cells is an intercellular substance called the middle *lamella* composed of *pectins*.

The cytoplasm is a jellylike substance or gel with clear fluid filled *vacuoles* not seen in animal cells. The cell nucleus is similar to the animal cells nucleus and will be described in detail later.

The two cells previously described are able to perform all the life activities of growth, division and metabolism, provided they are supported by the activities of *other* cells. Cheek lining and onion bulb scale cells are *specialised* cells for the purpose of performing a certain function. Isolated cells could not survive except in special carefully controlled *cultures* which provide foods, oxygen and warmth; such conditions are required for *transplant* organs and tissues awaiting use for transplant surgery.

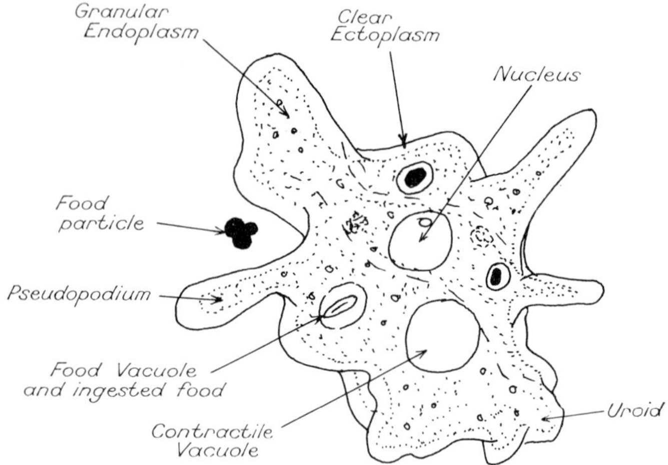

Fig. 1.3 Amoeba structure

A UNICELLULAR ANIMAL: AMOEBA

Amoeba is a unicellular animal found in ponds and ditches. The amoeba cell is composed of a clear jellylike *ectoplasm* and a granular *endoplasm*. Within the protoplasm are the nucleus and several vacuoles.

Life activities of amoeba

(1) **Irritability and movement.** Chemicals in the water may cause an amoeba to flow towards them if they are from food substances, whilst changes in light intensity, warmth, or pH can also cause movement in amoeba. Movement is a flowing of the cytoplasm into a projection called the *pseudopodium*, this *amoeboid* movement is seen in the white blood corpuscles of man.

(2) **Metabolism:** (a) *nutrition*. Food is *ingested*: the food particle together with water becoming engulfed by pseudopodia. This process is called *phagocytosis* and is performed also by white blood corpuscles which can remove solid bacteria particles from the blood as a function of body defence.

Insoluble foods are converted into soluble foods by *digestion* by means of *enzymes*. The digested soluble foods being absorbed by *diffusion* of the foods into the surrounding protoplasm.

Indigestible foods are *egested* through the ectoplasm as the amoeba moves forward.

(b) *Respiration* the process of liberating energy from digested food, is by means of oxygen which *diffuses* from the pond water into the cytoplasm, the energy released being used for growth, movement and body repair.

(c) *Excretion*. The by-products of respiration are either gases or soluble salts which accumulate in the water within the body cells. Water continually enters the body cell of the amoeba by *osmosis* through the *semipermeable* or selective membrane. Water accumulates inside the contractile vacuole which also dissolves the soluble excretory waste to finally emerge through the cell surface and enter the pond water.

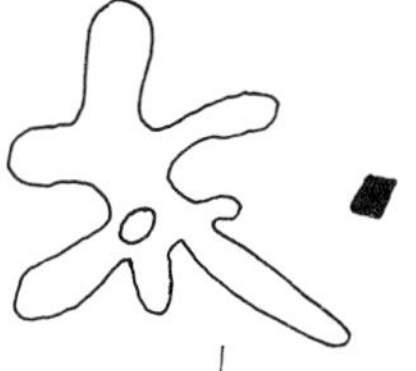

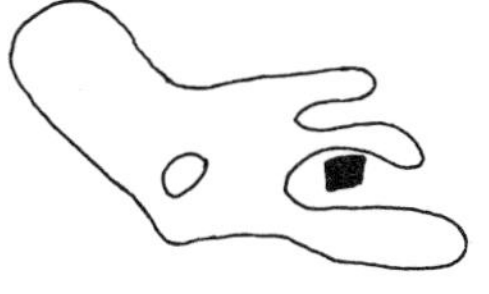

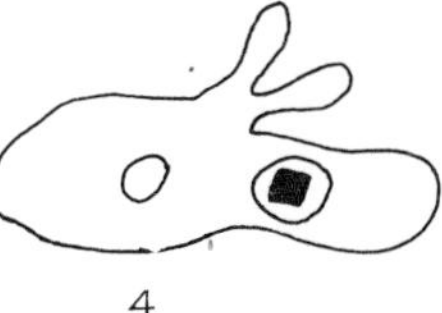

Fig. 1.4 Ingestion and movement in Amoeba

(3) **Growth.** Protein substances from the food are used in building the cell material which produces an increase in cell size.

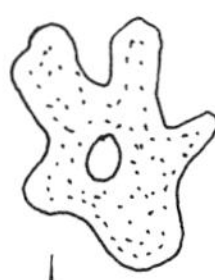

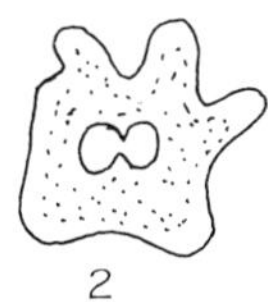

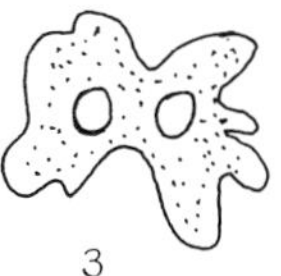

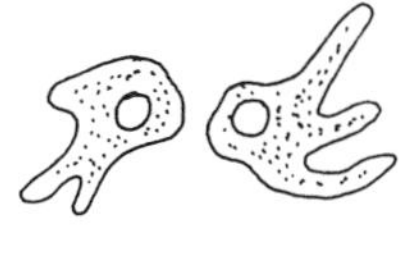

Fig. 1.5 Amoeba dividing

(4) **Reproduction.** An amoeba cell cannot continue to increase in size as this is limited by its *surface area* and *cell volume* relationships. When a certain cell size is reached the amoeba nucleus divides into two nuclei by a process of nuclear division called *mitosis.*

A UNICELLULAR PLANT: CHLAMYDOMONAS

Chlamydomonas is found in similar situations to the amoeba in ponds and ditches.

The cell wall is composed of a non-living material, *cellulose* found in all plants, with a *cell membrane* surrounding the cytoplasm containing the essential nucleus. In the cytoplasm is a large green coloured *chloroplast* containing *chlorophyll,* together with a *pyrenoid* body consisting of protein and granules of starch.

Life activities of Chlamydomonas

(1) **Irritability and movement.** A red *eye spot* or *stigma* appears to be sensitive to changes in light intensity, whilst the whiplike *flagella* propels the whole cell away or towards the light.

(2) **Metabolism:** (a) *nutrition.* Most green plants do not need to move about to obtain their food since they are able to manufacture their own food by *photosynthesis* from water and carbon dioxide using sunlight energy trapped by green chlorophyll.

Metabolism: (b) *excretion.* The by-products of respiration are carbon dioxide and water, the former can either diffuse into the surrounding water or be used for the purpose of photosynthesis.

Metabolism: (c) *respiration.* Oxygen is found dissolved in the pond water from which it passes by *diffusion* into the cell of Chlamydomonas.

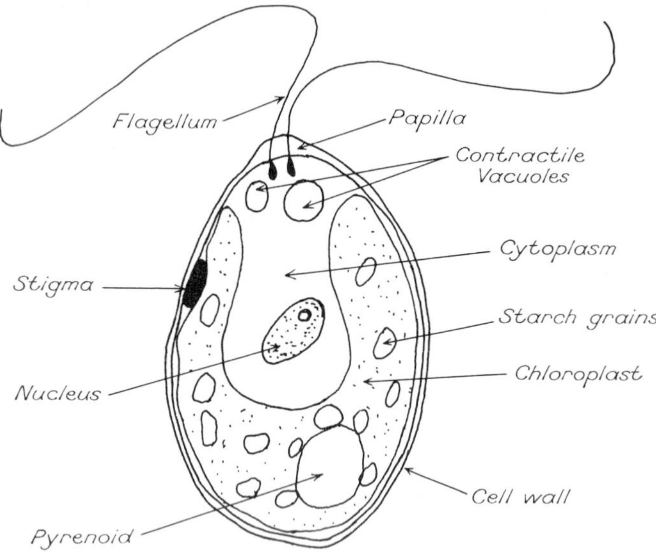

Fig. 1.6 Chlamydomonas structure

Metabolism: (d) *osmocontrol.* Since the Chlamydomonas cell is submerged in water there will be a movement of water by *osmosis* into the cell from the pond. Excess water will collect in the contractile vacuoles and be discharged or the water can be used in photosynthesis.

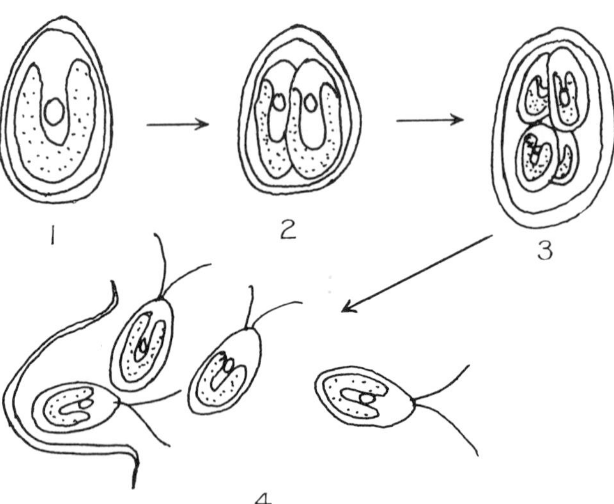

Fig. 1.7 Chlamydomonas asexual reproduction

(3) **Growth and reproduction.** When a certain growth size is reached the Chlamydomonas proceeds to divide by *nuclear division* to produce up to eight daughter cells that are liberated by the bursting of the mother cell wall. This process of reproduction involving one parent without sex is called *asexual reproduction.*

Differences between plants and animals

The main differences between plants and animals can be summarised as follows:

Plants	Animals	
Holophytic in which small molecules are changed into large molecules by *photosynthesis* Little, if any, digestive processes	*Holozoic* in which large molecules are used as food Complex digestive processes	**Nutrition**
Energy requirements are very small	Considerable energy needed together with special oxygen carriers in blood	**Respiration**
Simple chemical waste material, water and carbon dioxide easily removed	Complex and often toxic chemical waste made in large amounts needing rapid removal	**Excretion**
Fixed or *sessile* with little energy output (except Chlamydomonas)	Very active and mobile using large amounts of energy (except sponges)	**Movement**
Slow response to changes in surroundings, no special sense organs	*Rapid* response to changes in surroundings, through special sense organs	**Irritability**
Spreading and branched body *Cellulose* walls Rigid cell walls Vacuoles within cells Organelles include special chloroplasts	Compact streamlined bodies Cell walls of *lipoprotein* Flexible thin cell walls Non-vaculolated Chloroplast not found as cell organelles	**Structure**

DETAILED CELL STRUCTURES AND ORGANELLES

The very high powered *electron microscope* shows structures within the living cell called organelles not seen with the ordinary *light microscope*.

Many life activities such as respiration and photosynthesis are found to take place on the organelles. Figures 1.8–1.9 show the structure of typical plant and animal cells indicating the main types of organelles found in the cytoplasm. An important feature of the cell structure is the network of channels called the *endoplasmic reticulum* extending from the cell membrane to the nucleus, this shows the *porous* nature of the cell wall and explains how substances can freely enter the cells of plants and animals.

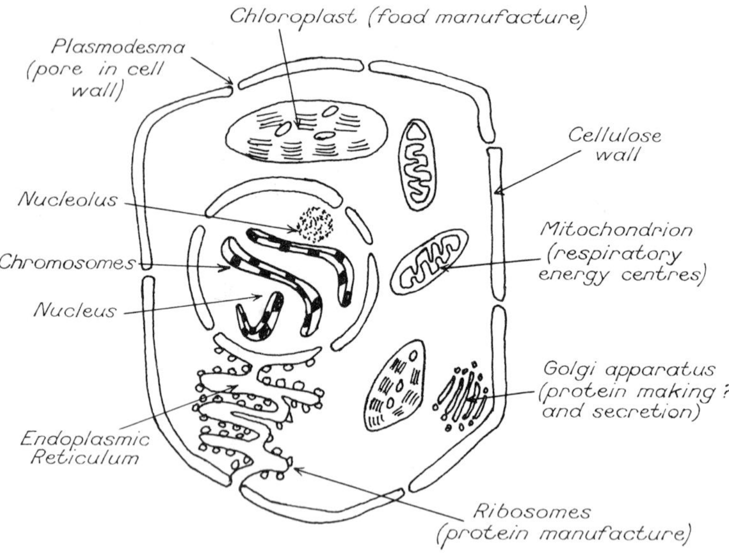

Fig. 1.8 Plant cell organelles

The following is a summary of the main plant and animal cell organelles and their function:

Organelle	**Function**
Mitochondria	Centres of respiration and release of energy
Lysosomes	Centres of digestion of complex foods into soluble foods
Chloroplasts	For food manufacture in plants by photosynthesis
Endoplasmic reticulum	Centre of protein manufacture for growth and repair on the *ribosomes*
Golgi apparatus	Secretory granules produced in most gland cells
Centrosomes	Concerned with cell division
Fibrils	Effect movement and change of cell shape
Cell inclusions	*Granules* of starch or glycogen
	Droplets of lipid oils
	Crystals of chemicals found in many plant cells, crystals are very rarely found in animal cells other than those which are diseased as in arthritis
	Colour pigments in plant and animal cells such as melanin in human skin

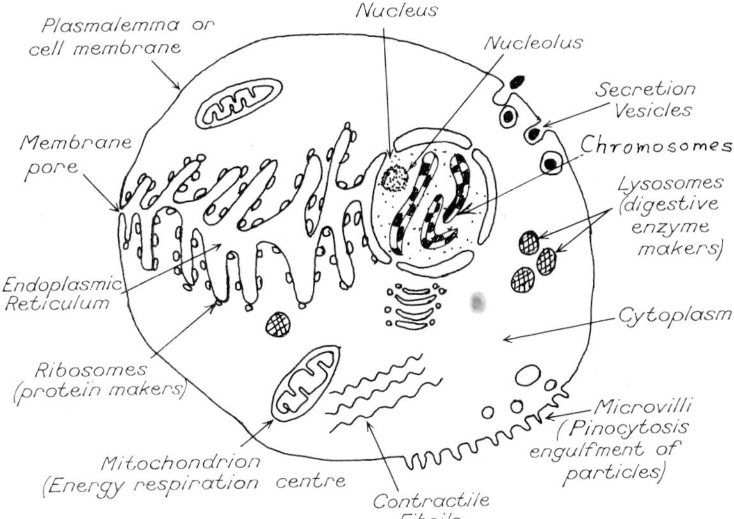

Fig. 1.9 Animal cell organelles

Pinocytosis. This is a process of engulfment which many animal cell walls are able to perform in a similar way to the phagocytosis of *amoeba* and white blood cells, by this method water and other substances are taken directly into the cell.

The main differences concerning the organelles of animal and plant cells are that animal cells include *lysosomes* for making digestive enzymes, *secretion vesicles*, and the *microvilli* concerned with engulfing particles in *pinocytosis*.

The animal cell has a membrane wall composed of *lipoprotein*, a compound of lipids and proteins, in contrast to the nonliving *cellulose* wall of plants.

SUMMARY OF CELL STRUCTURE

cell

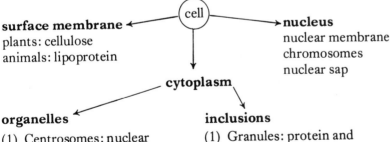

surface membrane
plants: cellulose
animals: lipoprotein

nucleus
nuclear membrane
chromosomes
nuclear sap

cytoplasm

organelles

(1) Centrosomes: nuclear division
(2) Mitochondria: energy
(3) Golgi bodies: secretion
(4) Lysosomes: digestion
(5) Fibrils: movement
(6) Endoplasmic reticulum: protein manufacture
(7) Chloroplasts: photosynthesis

inclusions

(1) Granules: protein and carbohydrates
(2) Pigments: in fruits and skin
(3) Crystals: mainly in plants

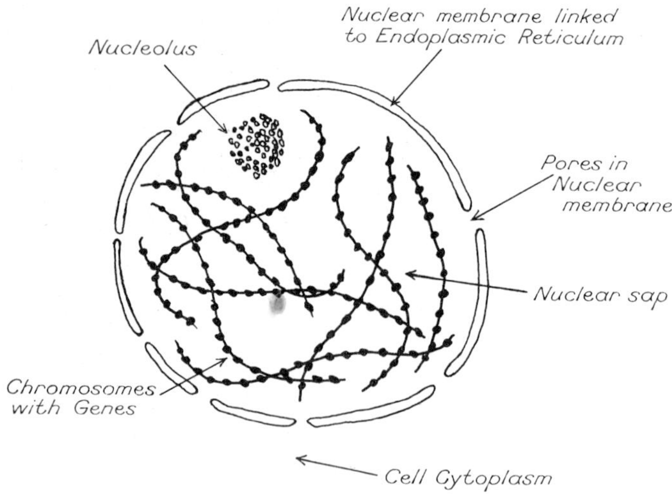

Plate 1 A karyogram photograph showing individual chromosomes from a human body cell nucleus (World Health Organisation) (facing page)

Fig. 1.10 Cell nucleus structure

CELL NUCLEUS STRUCTURES

The cell nucleus is surrounded by a nuclear *membrane* which has many pores allowing for the contents of the nucleus to connect freely with the cytoplasm. Cells which are attacked by cancer have very large nuclei almost filling the cell.

Chromosomes

These are the main component parts of a cell nucleus. They are found in pairs in body cell nuclei, there being a definite number of chromosome pairs for each kind of species of plant and animal:

frog	10 pairs	mouse	20 pairs
cow	30 pairs	chimpanzee	24 pairs
horse	33 pairs	dog	39 pairs
humans	23 pairs		

The photograph (Plate 1), called a *karyogram*, shows the chromosomes from a human cell nucleus; if counted 46 individual chromosomes will be found, making a total of 23 pairs.

Chromosome structure

A chromosome viewed through an electron microscope appears to be made up of two coiled *chromatids*. Along the lengths of the chromatids are tiny swellings, resembling beads in a necklace, called *genes*, or *chromomeres*. Genes made of deoxyribose nucleic acid (DNA) are responsible for the various features and characters that a person can inherit, such as eye colour, wavy or straight hair, etc.

A chromosome may have a *nucleolus* attached to it which is considered to be a reserve of protein substance related to ribose nucleic acid (RNA).

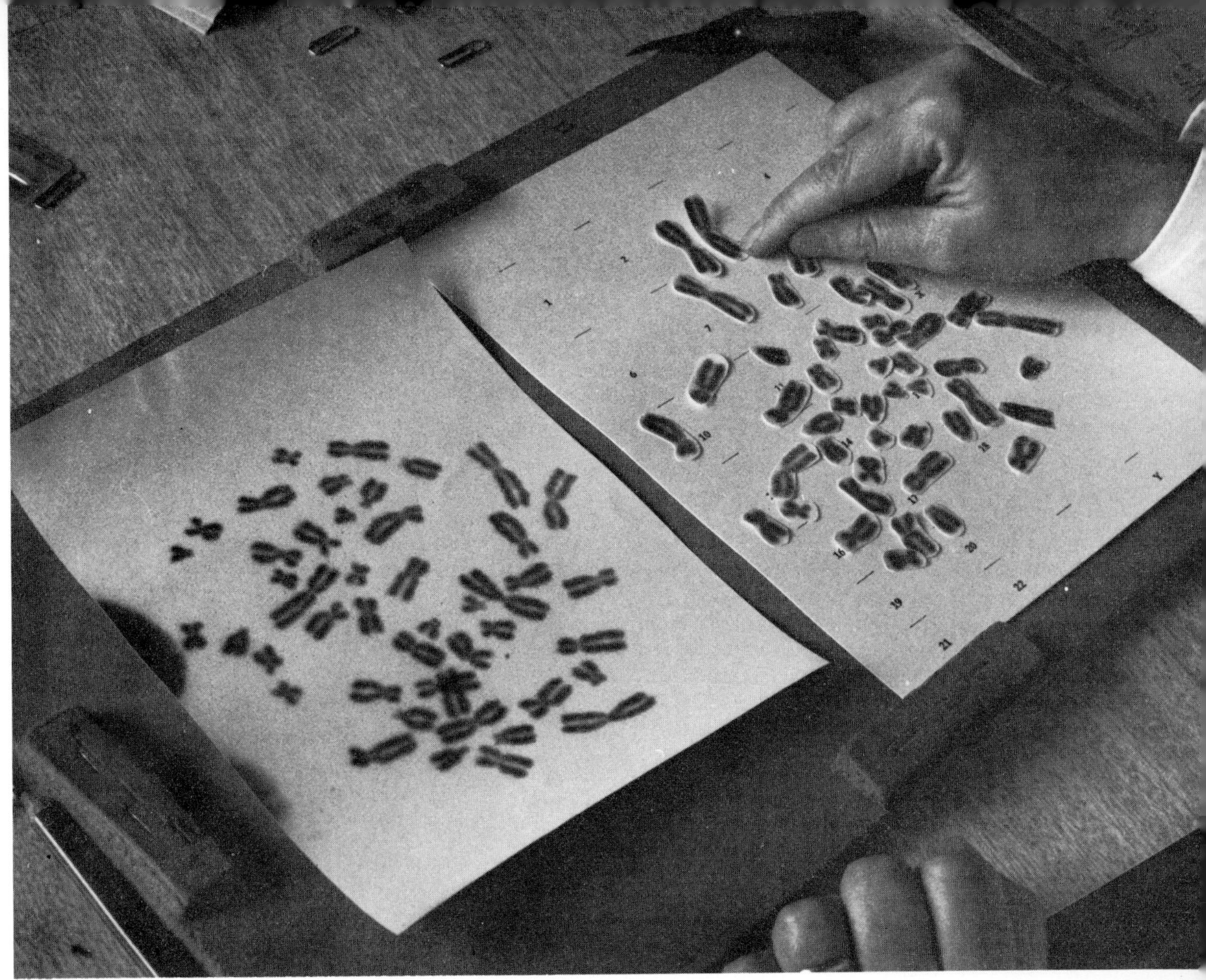

Division of cell nuclei

Plant and animal cells are either *body* (or *somatic*) cells or *reproductive* cells.

In *body* cells the nucleus divides by a process called *mitosis*, whilst the nuclei of reproductive cells, which produce the sexual gametes, divide by meiosis.

Mitosis

Nuclear division occurring in the body cells of plants and animals will result in cell division and finally in the *growth* of a plant or animal.

Figure 1.11 (p. 12) shows the different phases in the division of a body cell nucleus. The two daughter cells shown in D have the same number of chromosomes and structure as the parent cell nucleus in A. During the process of mitosis spindle fibres attach themselves to the *centromeres* and draw the *chromatids* to the *centrioles*, whilst a new wall forms between the new daughter nuclei.

A single fertilised human egg cell will produce millions of cells forming the complete human body by this process of mitotic nuclear division.

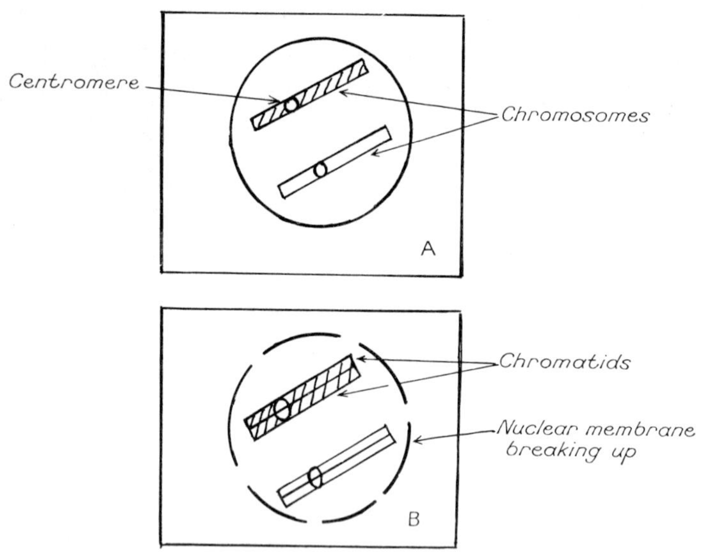

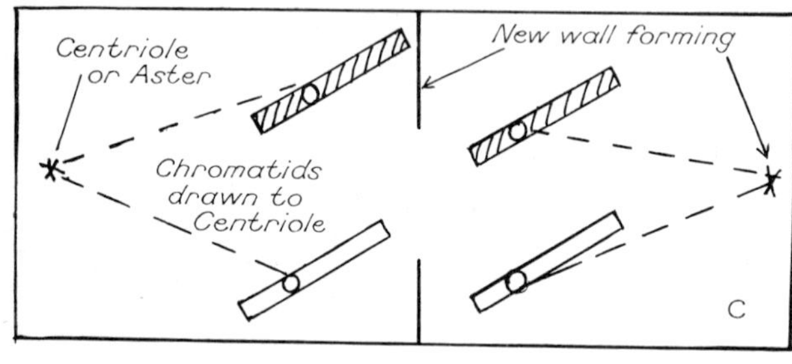

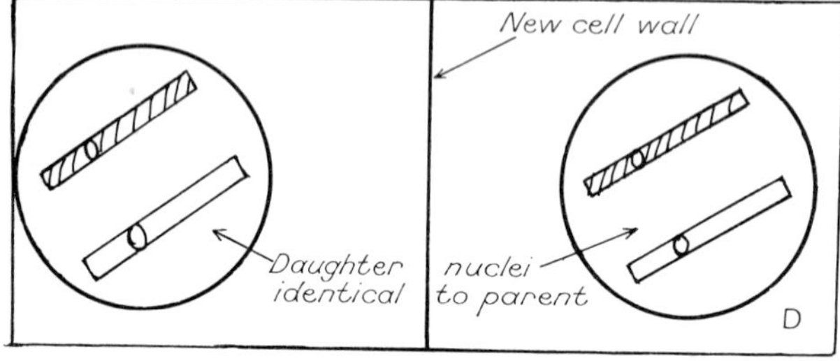

Fig. 1.11 Mitosis of cell nucleus

Antimitosis

Cancer invaded body cells divide at a rapid rate in a haphazard manner producing cancerous growths or ***neoplasms***. Certain cancer causing chemicals can affect mitosis in this way and produce neoplasms. Special drugs called ***antimitotic*** drugs are used to combat cancer growth by slowing down the process of nuclear division by mitosis, and consequently can reduce the growth (Plate 2).

Meiosis

Figure 1.12 summarises the complicated process of nucleus division taking place in the ***reproductive*** or *sex* cells of plants and animals. *One* cell nucleus divides to produce *four* cell nuclei, the original pair of chromosomes have been divided to give one chromosome to each daughter cell.

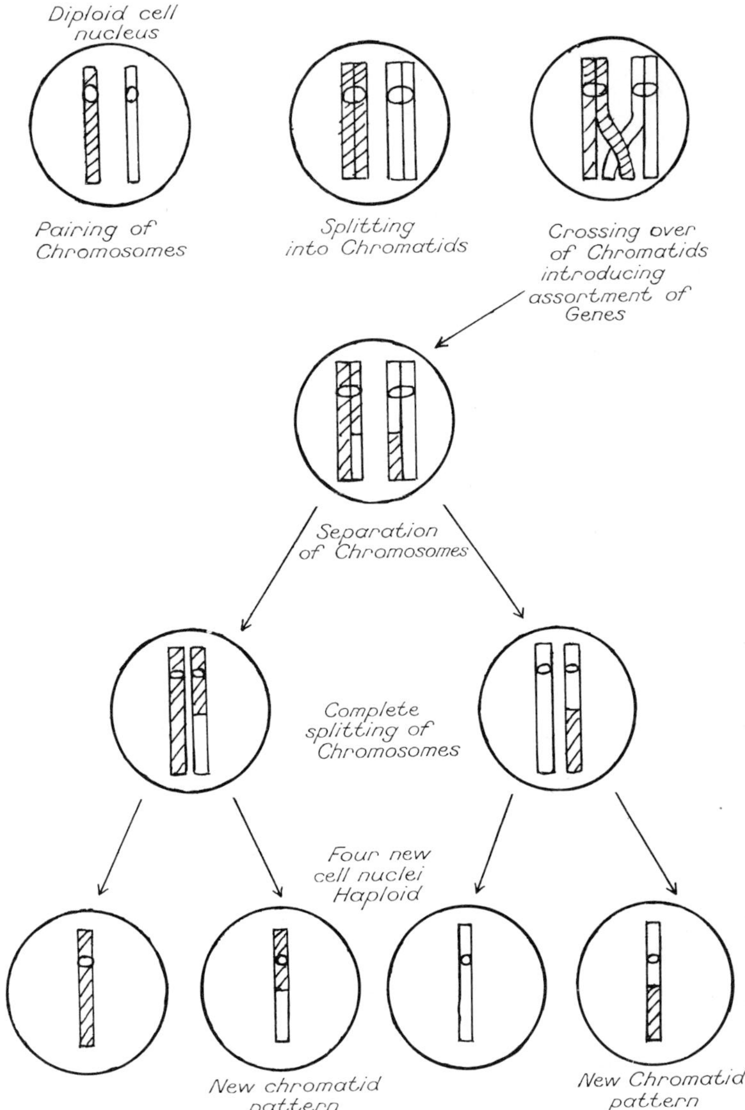

Fig. 1.12 Meiosis of cell nucleus

A very important feature of meiosis is the mixing and sorting of genes which occurs in a *crossing over* process, this will produce a new variety of chromosome pattern and consequently produce differences in the seeds of plants or the children of human beings.

Each *sex cell* or gamete of human cells has 23 chromosomes, instead of 46 chromosomes seen in the normal *body cells.* Sex cells having half the chromosome number are called *haploid.* Body cells with the normal number of chromosomes in the nucleus are called *diploid.*

Human body cells ⟶ 46 chromosomes ⟶ diploid
Human sex cells
(gametes) ⟶ 23 chromosomes ⟶ haploid

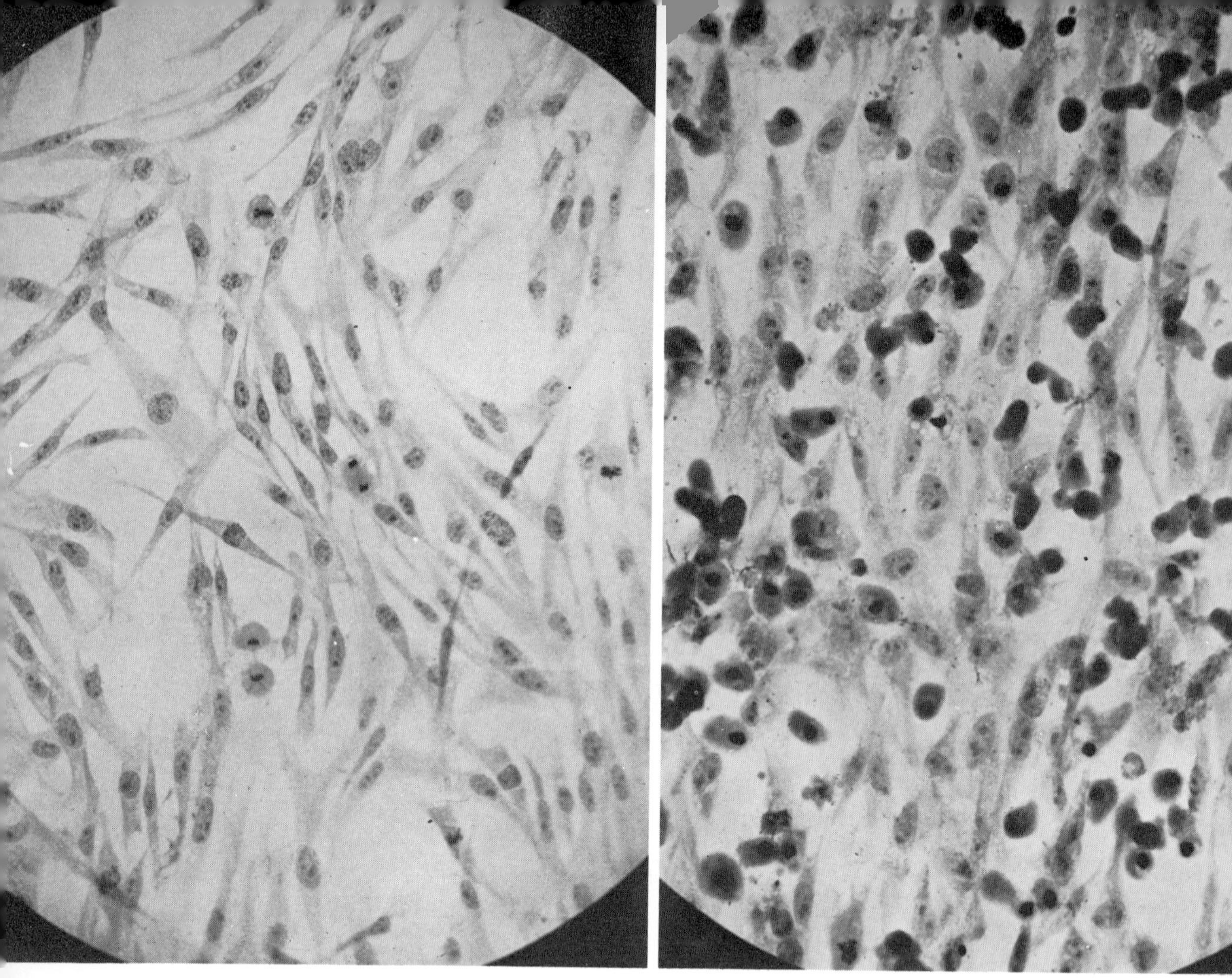

Plate 2 An antimitotic drug is shown having acted upon malignant cancer cells. Untreated cancer infected cells are shown on the left. The treated cells, shown on the right, have large darkly stained nuclei that have stopped dividing by mitosis (Sandoz Ltd)

When two gametes or haploid sex cells fertilise each other the two nuclei fuse together and form one nucleus of the *zygote* with a diploid number of chromosomes. This single nucleus proceeds to divide by mitosis and produce the new body cells of the plant of animal.

HAPLOID
Male sex gamete ↘ DIPLOID
Female sex gamete ↗ - - - - - - - - - - - - → zygote

Mature cells

When certain body cells reach maturity they no longer undergo mitotic division, such cells are those of bone, muscle, blood and nerves. Other cells can continue to divide by mitosis and include those found in skin, bone marrow, and gut lining.

SUMMARY

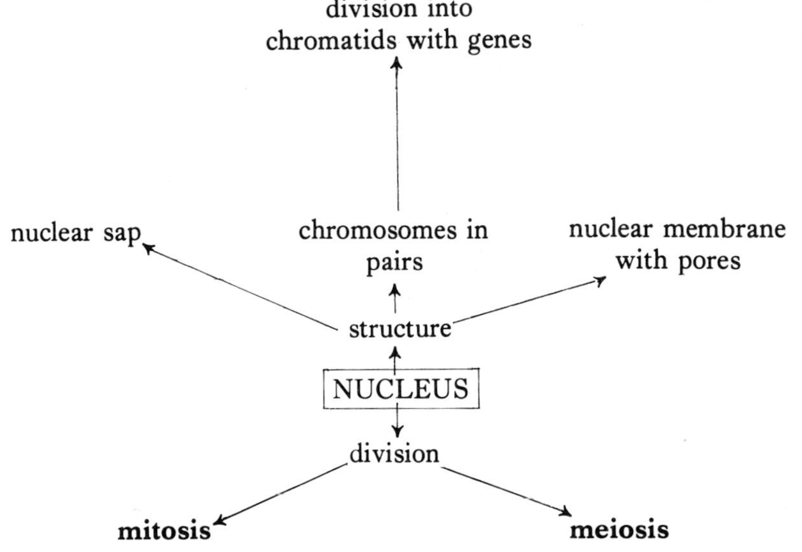

mitosis

(1) in body cells

(2) Daughter cell chromosome number as parent, i.e. diploid

(3) Daughter cells identical to parent cell

meiosis

(1) In reproductive sex cell or gametes

(2) Chromosome number is halved or is haploid

(3) Gene mixing in crossing over leads to variation

2 The Simple Animals

CLASSIFICATION OF ANIMALS

Animals are classified into groups or *phyla* of similar animals, based on the possession or certain structural features. The following summary is a simple outline classification of the animal kingdom:

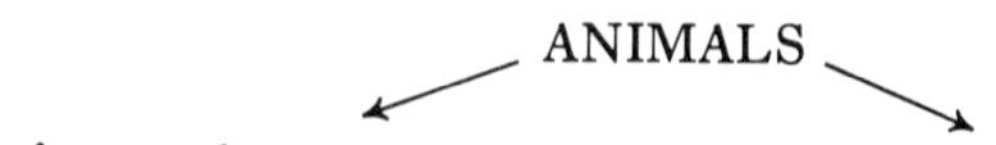

invertebrates
or spineless animals

(1) PROTOZOA, one celled, e.g. amoeba

(2) COELENTERATES, two layered multi-cellular, e.g. hydra

(3) HELMINTHES, flatworms, e.g. tapeworm

(4) ANNELIDA, segmented worms, e.g. earthworms

(5) ARTHROPODA, jointed limbed insects, crustaceans and spiders

(6) MOLLUSCA, shellfish, e.g. cockles and winkles

(7) ECHINODERMATA, spiny skinned, starfish

vertebrates
animals with a spine or backbone

(1) FISH, scaly skinned, gill breathers, e.g. cod and trout

(2) AMPHIBIA, moist skinned partly land and water living, e.g. toads and frogs

(3) REPTILES, scaly skinned, land living animals, e.g. snakes, lizards

(4) BIRDS, feathered, flying animals, some are flightless, e.g. penguins

(5) MAMMALS, furry or hairy vertebrates who suckle their young, e.g. man and the monkeys

PROTOZOA

Amoeba, already introduced as a simple unicellular animal, is also an example of an animal from the phylum Protozoa.

General features of protozoa

(1) Single celled animals performing all the life activities within *one cell.*
(2) Reproduction is by simple division or *fission.*

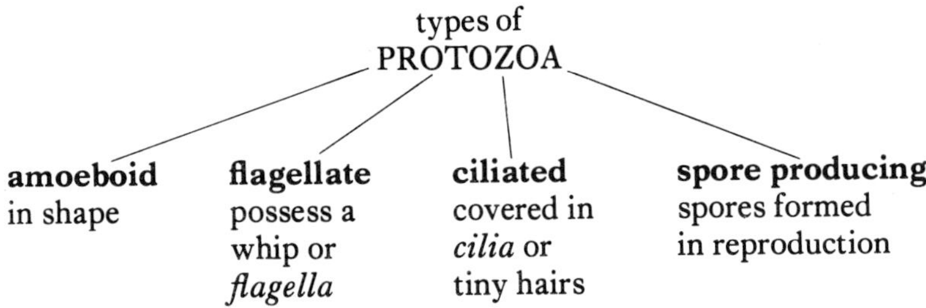

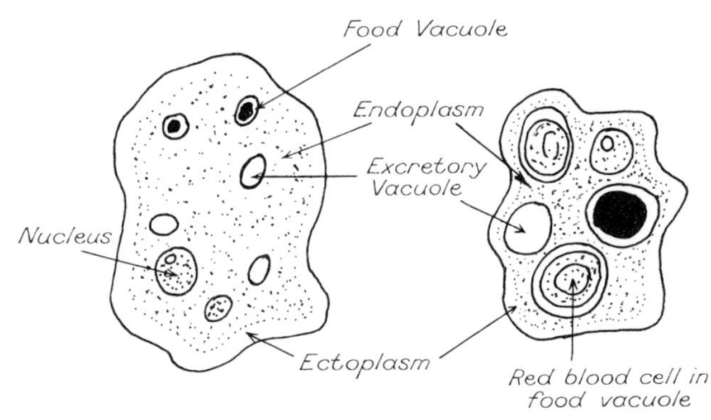

Fig. 2.1 Amoeboid protozoan parasites

AMOEBOID PROTOZOA

Two protozoan animals can be found living within the gut of man as *parasites*, both are called *Entamoeba.*

Entamoeba have cell structure similar to the amoeba, and engulf their food by *phagocytosis*. They reproduce by simple fission or mitotic division of the cell nucleus. *Entamoeba coli* is common in the large intestine or colon of man and is quite harmless.

Entamoeba histolytica found in the gut of man is harmful—causing stomach ache, amoebic dysentery and diarrhoea. Infection is by drinking contaminated water or eating unclean salads and vegetables. Houseflies transmit the *Entamoeba* from filth to food.

WHIP OR FLAGELLATE PROTOZOA

These protozoa have a definite shape through a firm cell wall together with one or more whiplike *flagella*. A single flagellum can move the animal cell through pond or sea water where many of these protozoa are found. They are also found as parasites in the body of man.

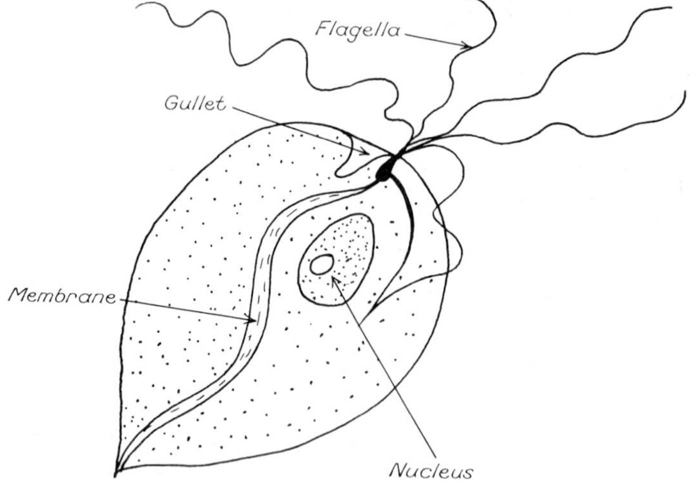

Fig. 2.2 Trichomonas vaginalis parasite

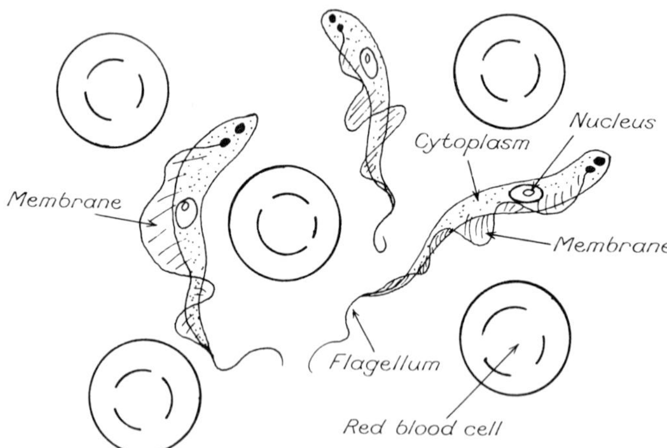

Fig. 2.3 Trypanosome cause of sleeping sickness

Flagellate protozoa found in man are shown in Fig. 2.2.–2.3. *Trichomonas vaginalis* is common in the vagina and excretory system of women, and to a lesser extent in men, where its presence may cause a discharge. Infection may take place during intercourse.

Trypanosomes are the protozoa found in the blood fluid of African animals and man causing the *sleeping sickness* disease. Infection and transmission is through the blood-sucking tsetse fly (Plate 3).

CILIATED OR HAIRY PROTOZOA

Ciliated or hairy protozoa are unicellular animals completely or partly covered in tiny hairlike *cilia*. The purpose of the cilia is to drive the animal cell through the fluid surroundings or to drive food towards a mouth region of the cell, by the flicking or beating action of the cilia.

Paramecium, shown in Fig. 2.4, is a common ciliated protozoan.

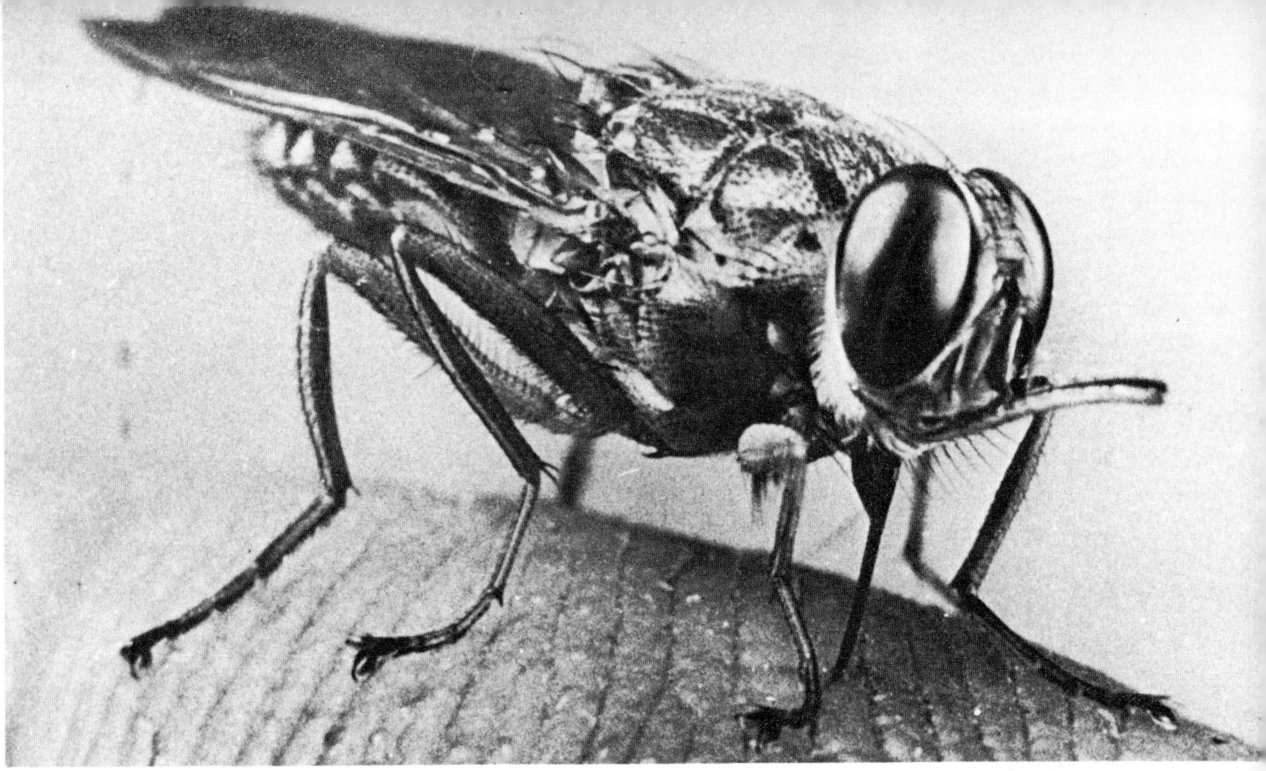

Plate 3 Tsetse fly which transmits the protozoan parasite of sleeping sickness (World Health Organisation)

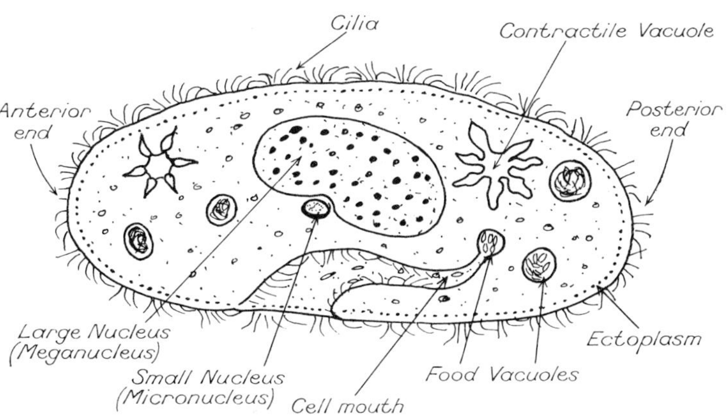

Fig. 2.4 Paramecium

SPORE PRODUCING PROTOZOA

Many animals can reproduce by producing great numbers of *spores* each of which has the ability to develop into the adult animal. The *malarial parasite* also called *Plasmodium* is a unicellular protozoa able to reproduce by spores. This protozoan is responsible for the tropical disease of *malaria* where the protozoan is found living *within* the blood cells of man, being transmitted to the blood from bites of the infected mosquito flies.

Figure 2.5 (p. 20) summarises the life cycle of the protozoan Plasmodium which causes malaria disease in man.

THE STINGING ANIMALS: COELENTERATES

In this phylum are included the tiny freshwater hydra and marine jellyfish, corals, and sea anemonies. Coelenterates and all other animals, except the Protozoa, are made up of more than one cell and are called *multicellular*. Animals with multicellular bodies can

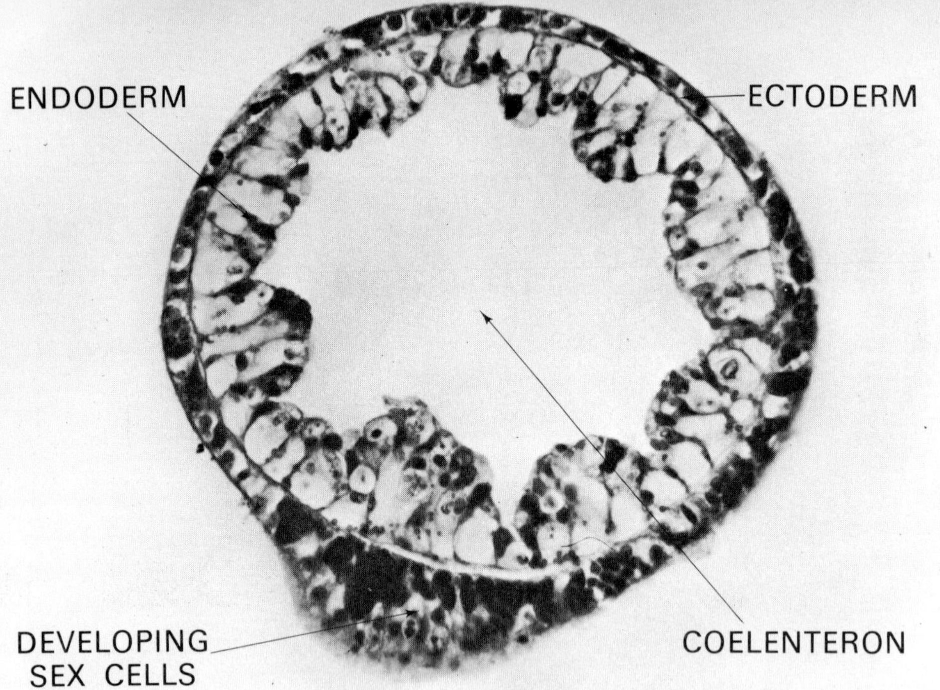

Plate 4 Transverse section of a Hydra body showing the ectoderm and endoderm cell body layers (Griffin Biological Laboratories)

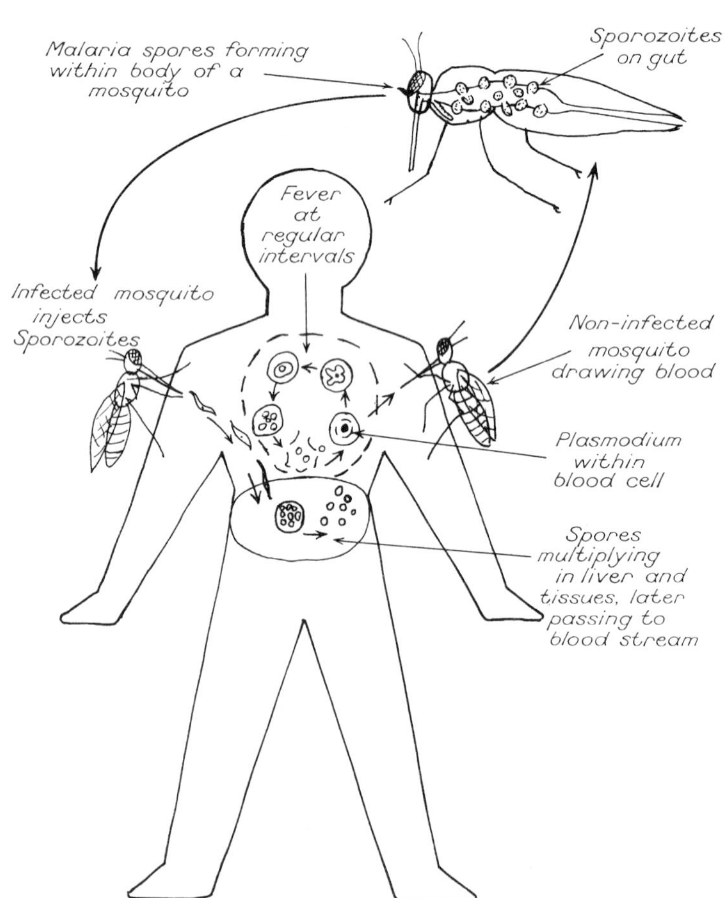

Fig. 2.5 Malaria parasite in man and mosquito

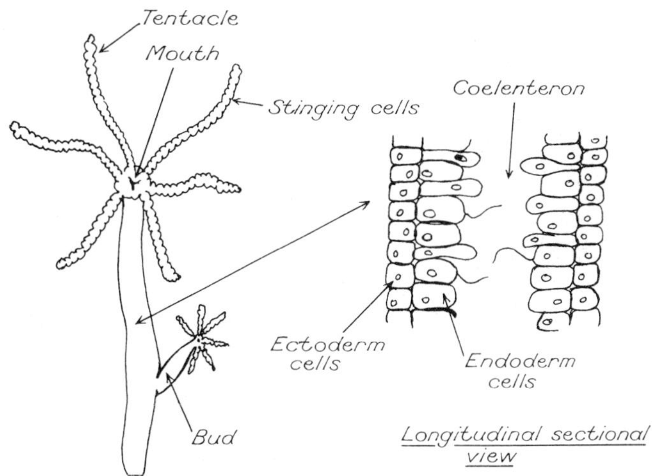

Fig. 2.6 Hydra

either arrange the cells in the form of a *ball*, with the cells heaped on top of each other and those cells in the centre unable to get a fair share of the food and oxygen, which must diffuse to them from the outside; or the cells can be arranged in *layers* of reasonable thickness which will allow easy diffusion of food and oxygen to reach all cells. Clearly the layered arrangement of cells is more efficient (Plate 4).

General features of coelenterates

(1) **The body cells** are arranged in two layers called the outer *ectoderm* and inner *endoderm*. The outer ectoderm cells are different kinds concerned with detecting changes in the surrounding water through *nerve* cells, or for movement by the muscle cells. Endoderm cells are mainly concerned with absorbing digested foods.

(2) **The two layered body wall** forms a large cavity called the *coelenteron*, which gives the name to this phylum of animals. This cavity serves mainly to digest food.

(3) **One opening, the mouth,** leads into the coelenteron, this must also function as the exit or anus.

(4) **Stinging cells** found on tentacles, needed to protect the animal, are an important feature of all coelenterates.

UNSEGMENTED WORMS OR HELMINTHES

Worms are mainly of two kinds *segmented*, like the earthworm, or *unsegmented* like the smooth threadworm.

The unsegmented helminth worms are of two kinds namely, flatworms and roundworms.

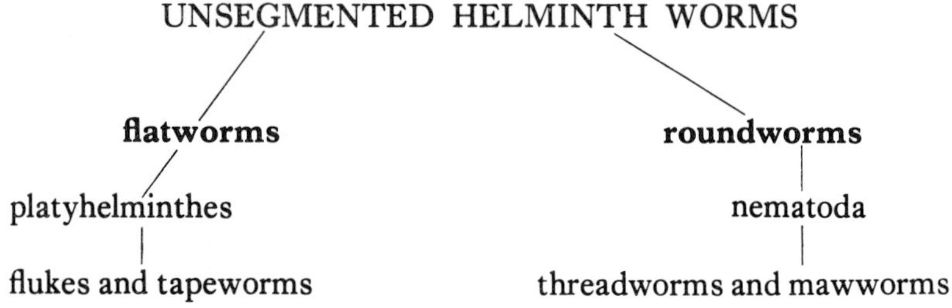

FLATWORMS

Two main kinds of flatworm are the *flukes* and *tapeworms*, each of which are multicellular animals in which the bodies are flattened from top to bottom. Flattening of the body gives the animal a *dorsal* back surface, and *ventral* belly surface, with right and left *lateral* sides. Animals which show a division of the body in this way are *bilaterally symmetrical.* If a *head* is present it is usually at the end which moves forward.

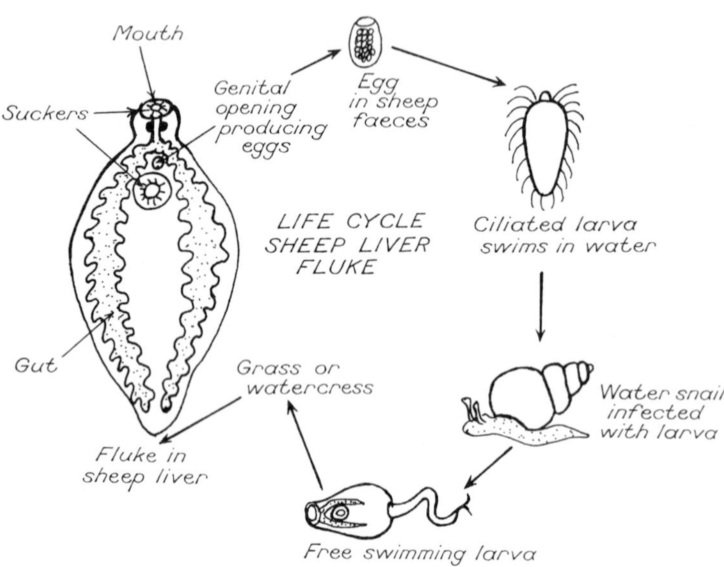

Fig. 2.7 Liver fluke

Flukes

Fluke worms are shaped like flattened privet leaves having suckers on one surface to anchor themselves to another living body, the *host* (Fig. 2.7). A mouth opens into a gut, the mouth being also the exit for faeces. Excretory organs discharge waste through an excretory opening. Fluke are parasites on other animals and in man are found in the following positions:

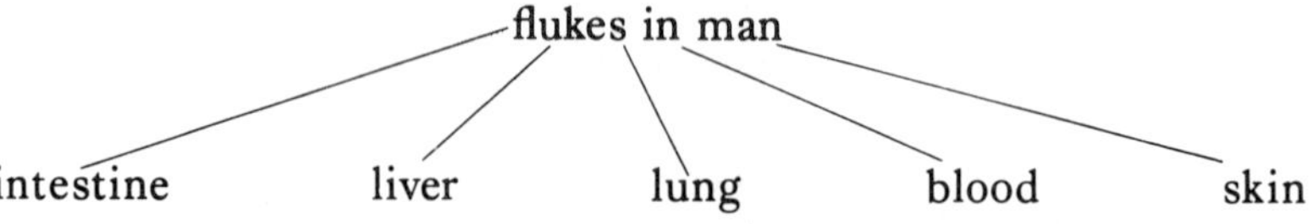

Fluke infections of man are found mainly in Africa, India, China, Far East, South America and Southern Europe. In Britain only the liver fluke is sometimes parasitic on man.

Liver fluke

Sheep and cattle are the most commonly infected animals with liver fluke, particularly those animals feeding in wet, waterlogged pastures.

Humans seldom become infected with liver fluke unless they have eaten contaminated watercress, or the rare case of eating raw infected liver.

All flukes have complicated life cycles summarised in Fig. 2.7.

Strict inspection of carcasses in abbatoirs by meat inspectors prevents the infection of man with the parasite.

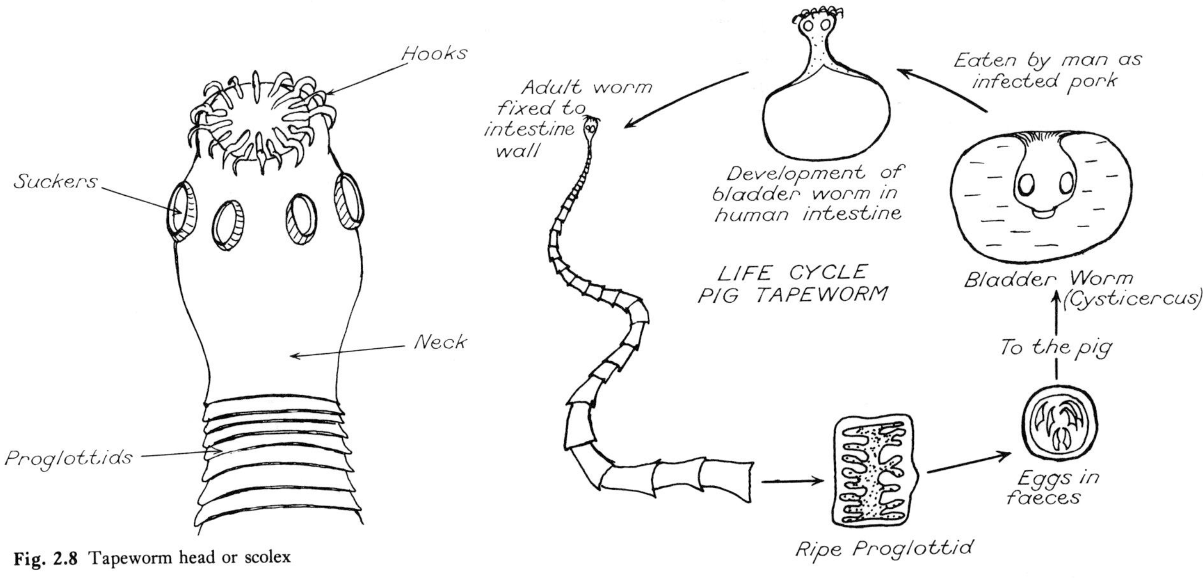

Fig. 2.8 Tapeworm head or scolex

Fig. 2.9 Life cycle of a pig tapeworm

Tapeworms (cestodes)

Adult tapeworms are flat worms with individual bodies called *proglottids* joined together to form a long tape of *strobila*, the tape has a head of *scolex* which can produce further proglottids by a process of *budding*. The adult tapeworm attaches itself to the gut wall of its host by means of hooks and suckers shown in Fig. 2.8.

Food enters the tapeworms from the gut in a pre-digested condition by direct absorption through the body surface; this feature shows how the animal is adapted to its parasitic way of life.

Life cycle of a pig tapeworm

Tapeworms have complicated lifecycles involving the *adult* tapeworm, and the *larval* bladderworm, summarised in Fig. 2.9.

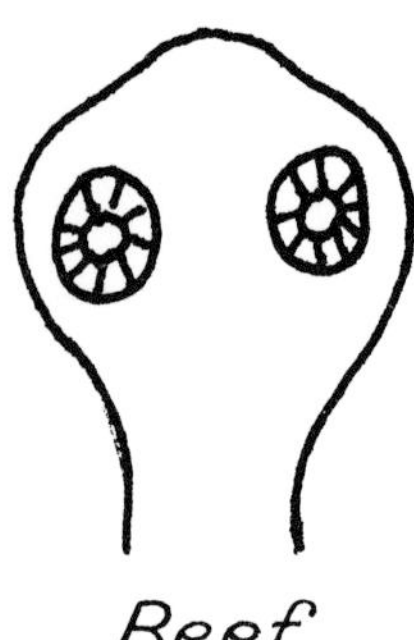

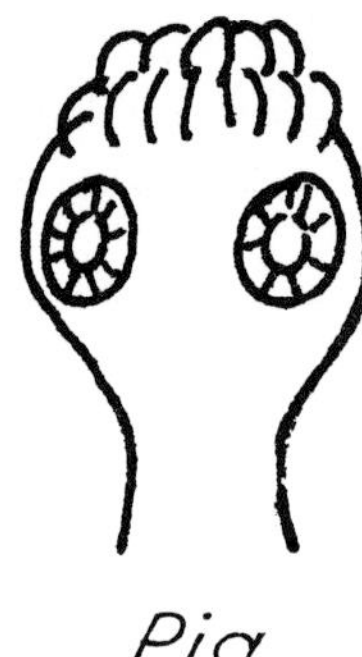

Fig. 2.10 Heads of different tapeworms

Adult tapeworms in man

These are the stages seen as the long *tape* with a head and proglottids in the small intestine.

Beef tapeworm is the most common of tapeworms found in man, being about 4 to 25 metres long, and introduced by eating raw or undercooked 'rare' beef infected with bladderworms.

Fish tapeworm is the next common tapeworm in Europe found in man, being 3 to 10 metres long, present as bladderworms in trout, salmon, pike and other *freshwater* fish which may be eaten raw or undercooked.

Pork tapeworm is one of the least common tapeworms in man, being 2 to 7 metres in length, and present as bladderworms in 'measly' pork infection being transmitted through eating raw pork or undercooked pork.

Dog tapeworm are very rare in man, being 10 to 70 centimetres long, the bladderworm being carried by cat or dog *fleas* which may be accidentally eaten by man!

Strict inspection of meat by meat inspectors reduces the incidence of adult tapeworm infection in man. Thorough cooking and the use of efficient sewage disposal systems reduce the incidence of the disease still further.

Various treatments of adult tapeworm infection include the use of drugs to displace the tapeworm this is followed by careful examination of the faeces to confirm the expulsion of the tapeworm head (Fig. 2.10).

Larval bladderworms in man

The *egg*, entering the host, can develop into a *bladderworm* or *cysticercus* larva tapeworm in different organs of the body. In man the organ to become infected with bladderworm can include the lungs, liver, kidneys, brain or muscles, such infections are very serious and fortunately rare.

Pork bladderworm can be found in the human brain and muscles of the body, infection is from water and food contaminated by human faeces carrying pork tapeworm proglottid eggs. Uncleaned hands may be the cause of self-infection.

Dog bladderworm form *hydatid* cysts in the liver, lungs or kidney of man. The eggs enter humans from close contact with dogs, particularly sheep dogs, food may be contaminated by infected dog faeces, or close petting of pet dogs.

Removal of bladderworm larval tapeworm in man is achieved by surgical methods.

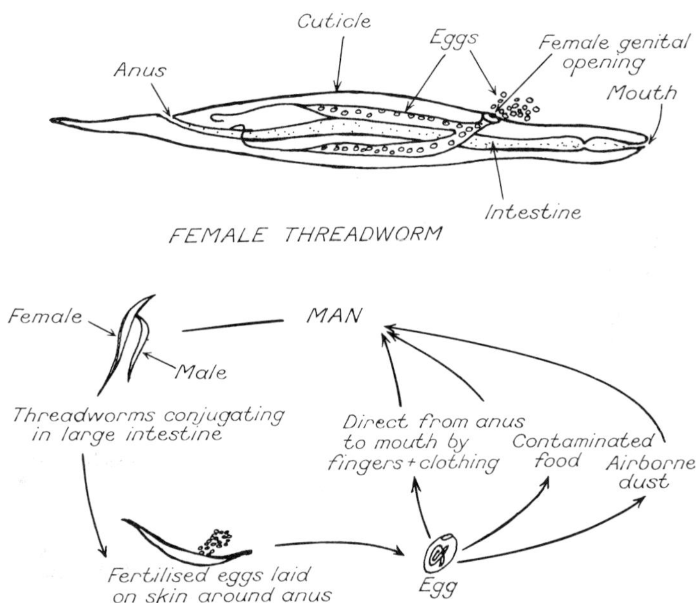

Fig. 2.11 Life cycle of a threadworm

Roundworms (nematodes)

Roundworms have long thin, round or cylindrical, shaped bodies tapered at both ends. The outer skin is covered with a tough outer coat called a *cuticle*, this is resistant to the digestive juices of the host animal.

The alimentary canal is a straight tube commencing at the mouth and ending as the *anus*, being lined throughout by *endoderm*.

Male and *female* roundworms are found, in comparison to the *hermaphrodite* tapeworm. Sexual reproduction takes place between the opposite sexes resulting in the female roundworm laying many fertilised eggs.

The fertilised eggs hatch into tiny roundworms called *juveniles* shown in the life cycle summary (Fig. 2.11) which grow into adult worms after several *moults* in which the outer cuticle is cast off.

Roundworms are responsible for many parasitic diseases in man, from the common *threadworm* disease of children to the rare *trichina worm* disease.

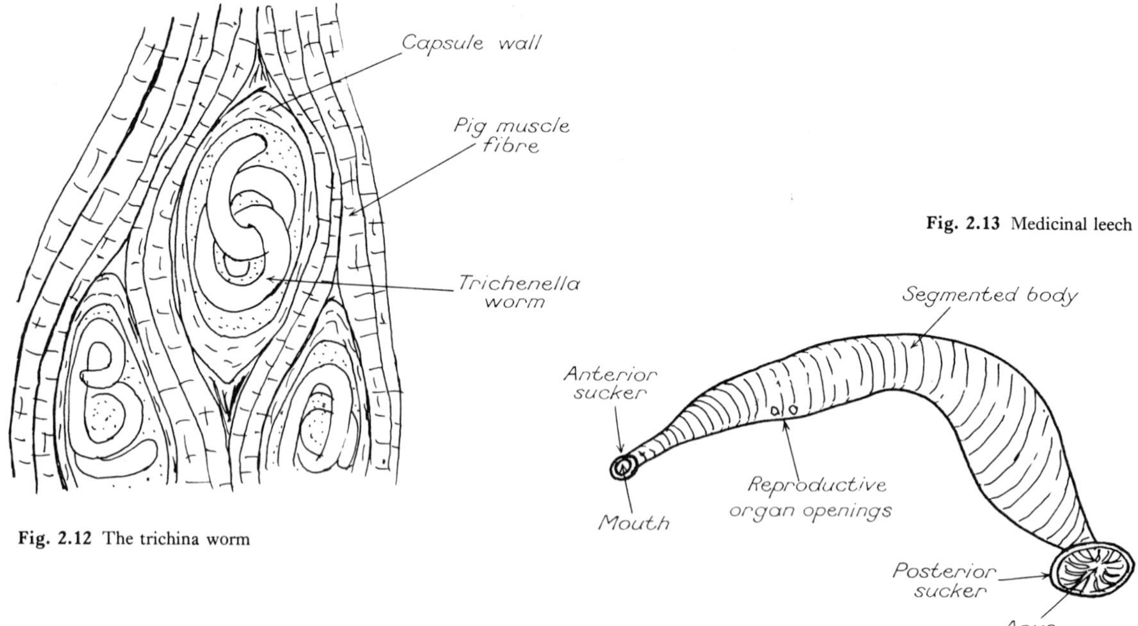

Fig. 2.12 The trichina worm

Fig. 2.13 Medicinal leech

The following summarises some of the nematode diseases of man.

Name and size	Location in man	Method of infection
Threadworm or pinworm 2–10 mm	In children from stomach to rectum. Eggs laid at night on anus causing itching	Eggs in contaminated water and food. Self infection from fingers. Airborne dust and soiled clothing
Ascaris or mawworm 15–35 mm	Adults in intestine juveniles in lungs	Soil contaminated water and food or on hands
Trichina worm 1.5–5 mm	Juveniles encyst in body muscles causing *trichinosis* with muscle pain and breathing difficulty	Parasite of rats, passed to pigs. Infected pork eaten raw or undercooked infects man
Whipworm 2–5 cm	In the large intestine	Soiled food or water
Hookworm 5–13 mm	Tropical and some European countries, in the small intestine	Juveniles *bore through the skin* of the feet or in contaminated water

SEGMENTED WORMS OR ANNELIDS

Segmented worms, include the earthworm and leech, whose bodies show a division into *segments* seen outside and also inside the body on dissection. The internal segments are seen as *compartments* with dividing walls or *septa.*

External features

Earthworms have an outer *cuticle* covering the ectoderm with an anterior mouth and posterior anus. *Bristles* are found on each segment as pairs serving in movement of the animal. The thickened *saddle* found on segment 32 is used for producing egg cases.

Leeches also show segmentation but are without bristles, instead they have *suckers,* one around the mouth and one at the posterior end to serve in movement. A saddle is found on segments 9–11.

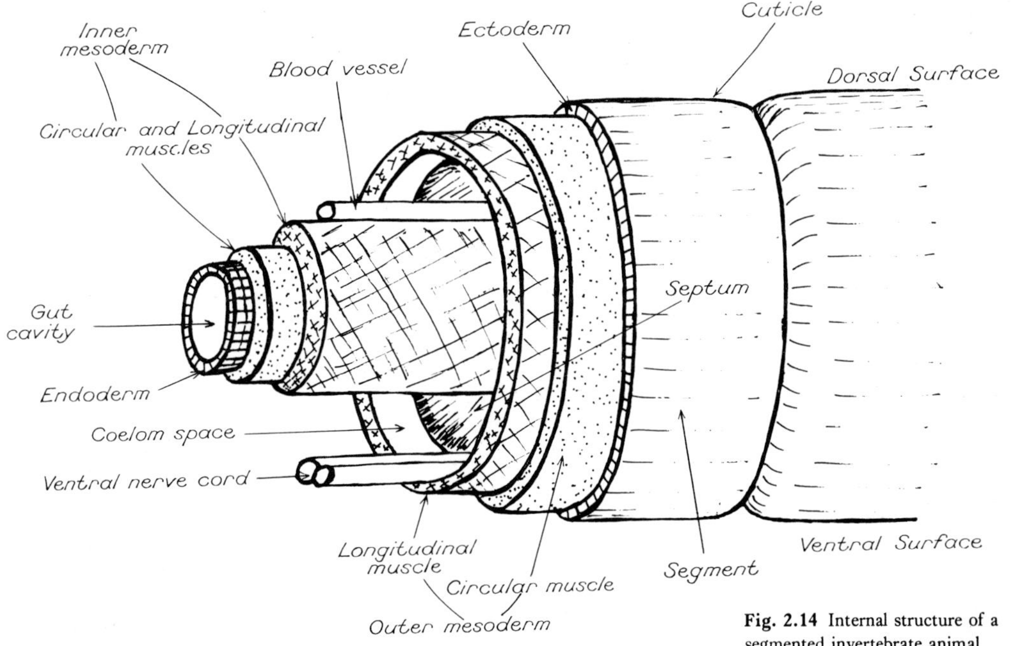

Fig. 2.14 Internal structure of a segmented invertebrate animal

Body structure

Earthworm and leech are multicellular animals in which the body cells are arranged in *three* main layers. An outer *ectoderm,* inner *endoderm* and a middle layer the *mesoderm.* This is an additional body layer compared to the two layered body structure of hydra.

The middle layer or mesoderm shows a space full of fluid called the *coelom,* this separates the mesoderm into outer and inner mesoderm layers, which are mainly muscle layers. The three layered structure of the annelid worms is seen in a *transverse* section of an earthworm shown on prepared microscope slides viewed through a hand lens. (See Fig. 2.15 overleaf.)

Animals including man having this three layered structure, together with the coelom, are called *triploblastic* and *coelomate.*

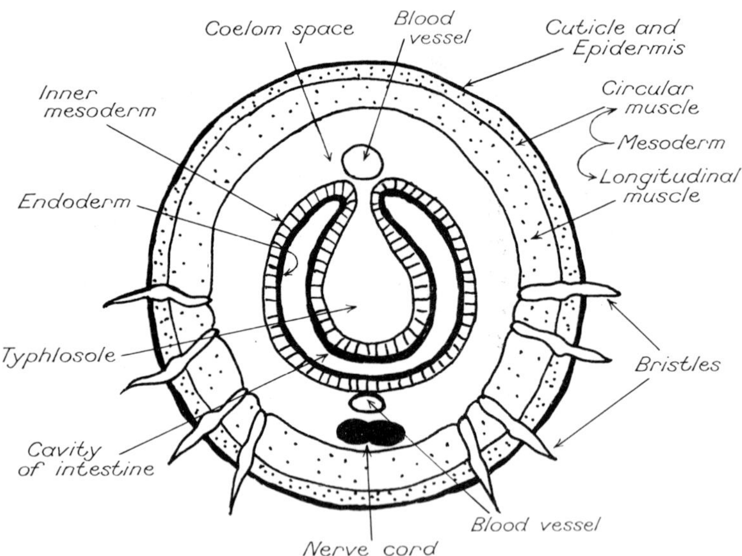

Fig. 2.15 Transverse section of an earthworm

Nutrition

The complex insoluble foods are taken in the mouth with large amounts of soil and digested by digestive juices containing *enzymes* secreted by the endoderm cells lining the long tubular *intestine*. The endoderm cells of the intestine serve also to *absorb* the simple soluble digested food. Absorption will take place quickly over a large *surface area*. In the earthworm the total surface area is increased by an infolding, or *typhlosole*, of the intestine.

Circulation

Circulation systems are essential in multicellular animals to provide all the living cells with food and oxygen and to remove excretory waste products.

The blood of annelid worms is a fluid *without* blood cells. Soluble food enters the blood vessels from the intestine and is driven forward along the blood vessels by the action of *muscles in the vessel walls*, there is *no heart* to function as a pump.

Respiration

Oxygen from the air passes by *diffusion* through the moist thin ectoderm and into the blood vessels of the outer mesoderm. Oxygen combines with a red pigment in the worm's blood fluid and the oxygen is carried to all cells requiring oxygen.

Excretion

Excretory substances collect in the *coelomic fluid* of each segment, this is removed by special excretory organs called *nephridia* which are very simple kidneys, each of which has a good supply of blood bringing soluble waste for removal. The nephridia are fixed into the segment septum and discharges its waste onto the outer skin surface, through a pore.

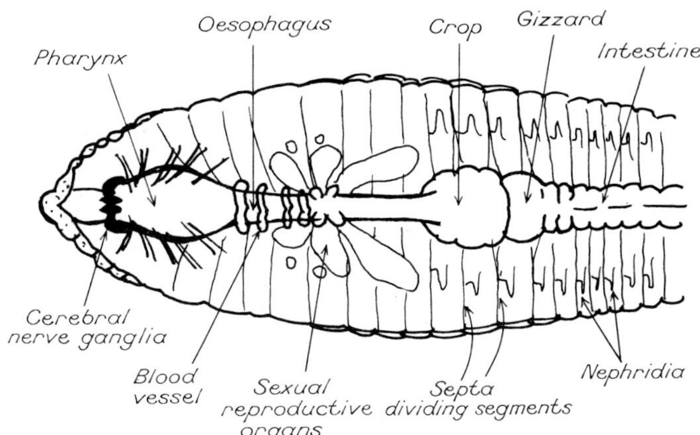

Fig. 2.16 Internal structure of an earthworm

Reproduction

Annelid worms are *hermaphrodite* in having both male and female reproductive organs, called the male *testes* and female *ovaries*. The sex *gametes*, *spermatozoa* and *ova* are shed into the coelomic fluid.

Sperm is stored in the *vesiculae seminales* until required for the process of *cross fertilisation* when it fertilises the ova of another worm.

Nervous system

Activities between the many body cells is *coordinated* by the nerves. Worms are sensitive to the stimulus of light and to touch through *sensory* cells in the ectoderm, which are linked with the *nerve cells*. The nerve cells of worms are collected together in a *ventral* position as pairs of *ganglia*. The ventral position of the nerve cord is possibly due to the continual stimulation of the underside or ventral surface as it crawls over the soil. The head end of the worm which moves forwards and consequently receives most stimuli, developes *cerebral* or *brain* ganglia.

Movement

Two kinds of muscles are found in the earthworm both formed from the *mesoderm*, *longitudinal* and *circular* muscles which always occur together. Circular muscles contract to produce a small opening and relax to form a large opening, whilst longitudinal muscles contract or shorten and relax to lengthen. When a circular muscle contracts the longitudinal muscle relaxes and vice versa, such pairs of muscles which work in opposition to each other are called *antagonistic pairs*.

Relaxation of the longitudinal muscles with *contraction* of the body wall circular muscles cause the worm to *elongate*, whilst *contraction* of the longitudinal muscles with *relaxation* of the circular muscles in the body wall cause the worm to *shorten*. By digging its bristles into the soil surface the worm is able to move in waves of body contraction and relaxation.

3 *Jointed Limbed Invertebrates: Arthropoda*

Almost 75% of all known animals are members of this phylum, the phylum being composed of the following classes:

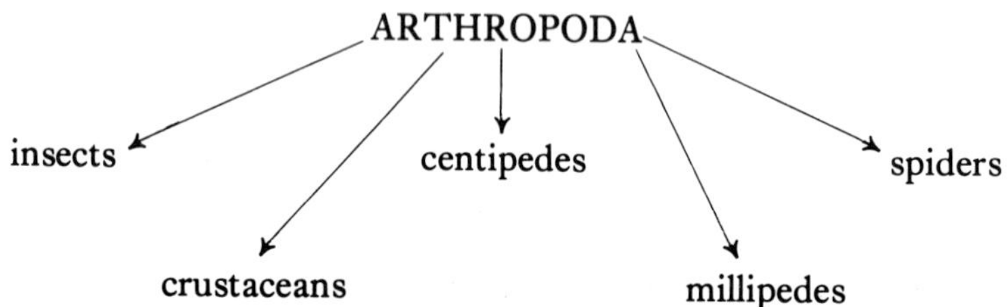

FEATURES OF THE ARTHROPOD ANIMALS

(1) **An exoskeleton** composed of a tough flexible substance called *chitin* sometimes hardened with calcium salts forms an outer protective shell to the body, thus protecting the soft internal organs.

(2) **Segmentation** is seen externally.

(3) **Appendages** called jointed limbs are attached to the body.

(4) **A heart** is present to circulate the blood.

INSECTS

Most of the arthropod animals are insects, a few of which affect the health of man by causing or transmitting certain diseases. Insects are classified into the following groups:

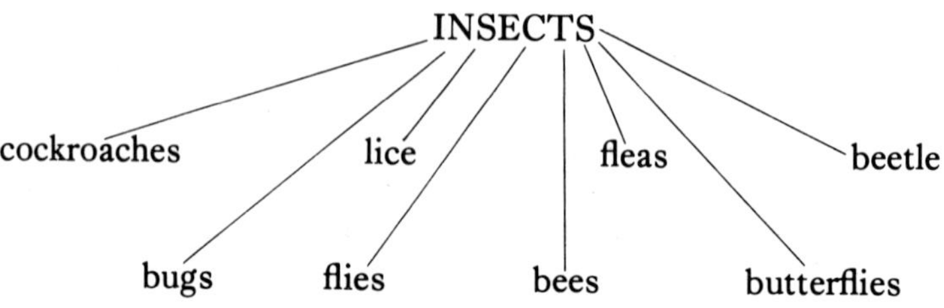

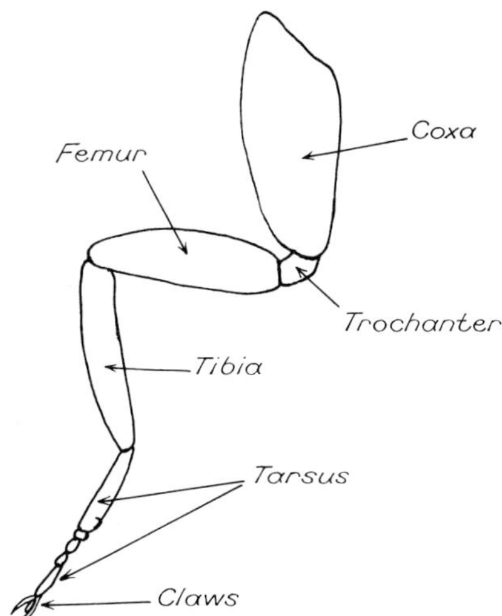

Fig. 3.1 Jointed limb of an arthropod insect

FLIES (DIPTERA)

Flies are the *two winged* insects and include: housefly, bluebottle, mosquito, horsefly and gnats.

The housefly

This insect has the typical body structure in which it is divided into three parts, head, thorax and abdomen.

(a) **The head** has compound eyes, and the *antennae* are concerned with detecting smell. The mouth is a sucking organ or *proboscis*. The appendages called the *maxillary palps* are concerned with examining food by taste and touch.

(b) **The thorax** is composed of three large segments divided into dorsal and ventral plates. A single pair of *wings* is attached to the upper dorsal plates. Three pairs of jointed legs each consisting of parts called the *femur*, *tibia* and *tarsus* terminating in an adhesive pad between claws, the pad allows the fly to walk clinging to a ceiling upside down.

(c) **The abdomen** is divided into eight segments with flexible chitin between the segments and joints. The whole body is covered in sensory hairs called *macrochaetae*.

Since the body of a young or immature insect is surrounded by the restricting exoskeleton it is cast off in a series of moults to allow it to increase in size, this moulting is called *ecdysis*.

Nutrition

Flies feed on filth, faeces, rotting refuse and also human food; they secrete saliva from their salivary glands to convert the food into a soluble form, or alternatively vomit from their own gut to digest the food: this is seen as 'flyspots'. Flies therefore are *mechanical vectors* in transmitting disease to man, through the microbes carried on the body hairs or in the saliva or vomit and faeces deposited on human food.

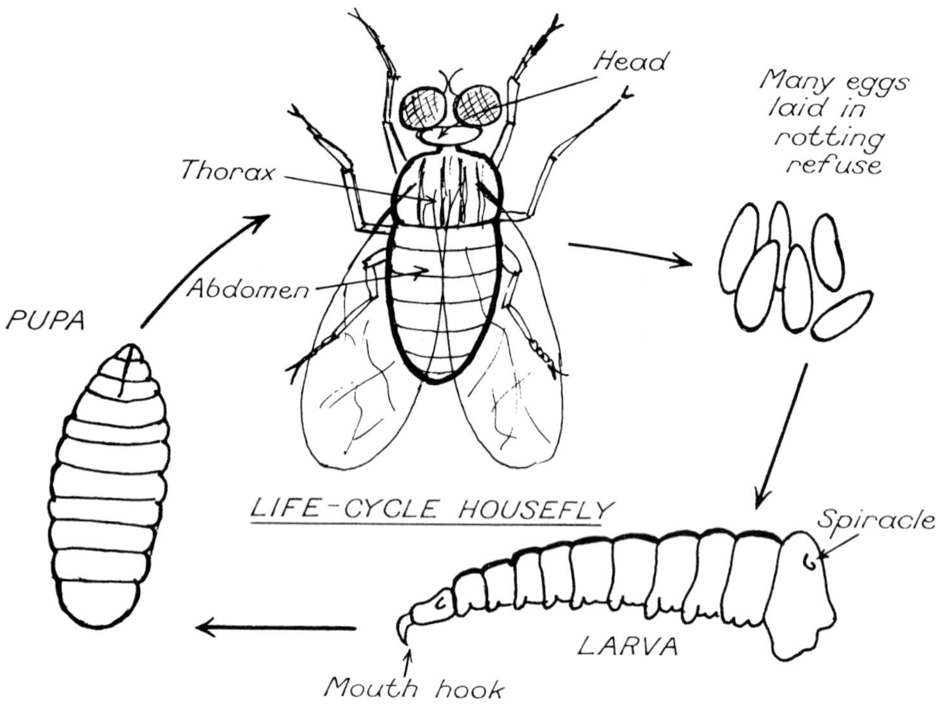

Fig. 3.2 Housefly life cycle

The following shows how the housefly acts as a mechanical vector of several diseases.

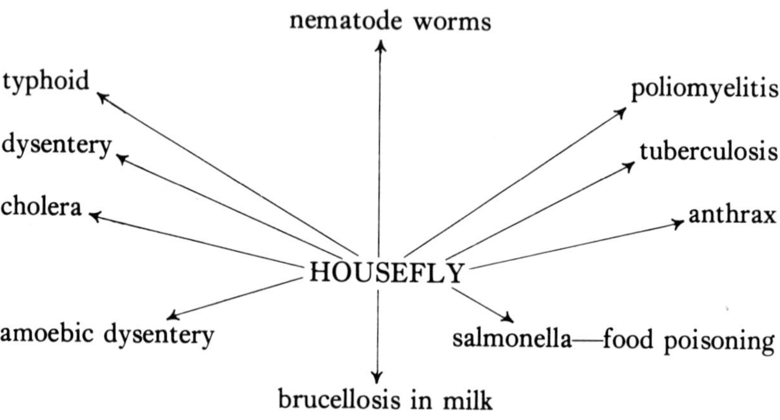

Life cycle

The distinctly separate male and female flies reproduce sexually to produce great numbers of offspring, the life cycle is summarised in Fig. 3.2. Insect development is called a *complete metamorphosis* if it

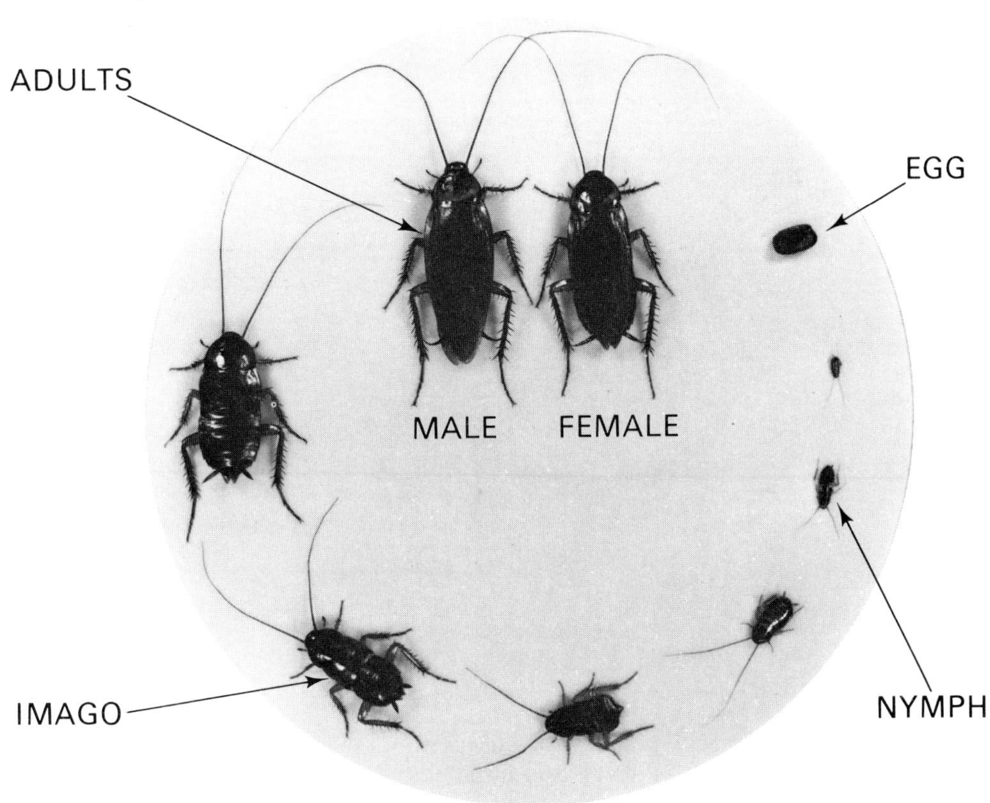

Plate 5 Incomplete metamorphosis seen in the Cockroach life cycle (Rentokil)

proceeds from egg → larve → pupa → imago; this development is seen in fleas, bugs, wasps, and most flies.

Life activities of insects

Incomplete metamorphosis proceeds from egg → nymph → imago, without a larva or pupa stage; this is seen in the cockroach.

Circulation

Blood fluid fills the blood spaces surrounding internal organs, the fluid is circulated by a *tubular heart*, the digested food being carried to the body cells by the blood fluid (Fig. 3.4).

Respiration

The thick, dry body wall of chitin is unable to allow oxygen to diffuse into the body, air must therefore be taken directly to the body cells by a system of tubules called *tracheae* which are connected to the exterior through openings called *spiracles*. Movement of the body wall by muscular action draws air into the tracheae tubules.

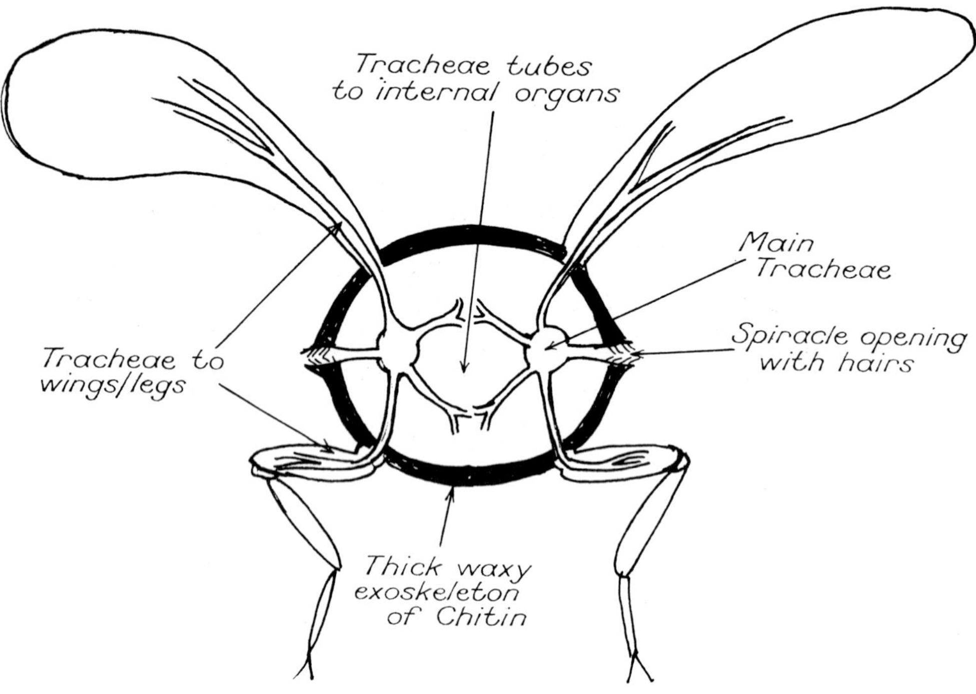

Fig. 3.3 Respiratory system of an insect

Excretion

Excretory waste is collected from the blood fluid by special ***malpighian tubules*** that pass the waste into the gut to be removed along with the faecal waste. In order that the insect conserves its water supply, the malpighian tubules withdraw water from excretory material so forming a solid excretory product (Fig. 3.4).

Nervous system

A highly developed system is found consisting of a ventral nerve cord together with special sensory organs of the eye, and antennae.

Movement

Antagonistic action of muscle pairs *within* the exoskeleton result in limb movements; as one muscle contracts the other relaxes this causes movement of the limb section by the muscle pulling on muscle *ropes* acting through flexible joints.

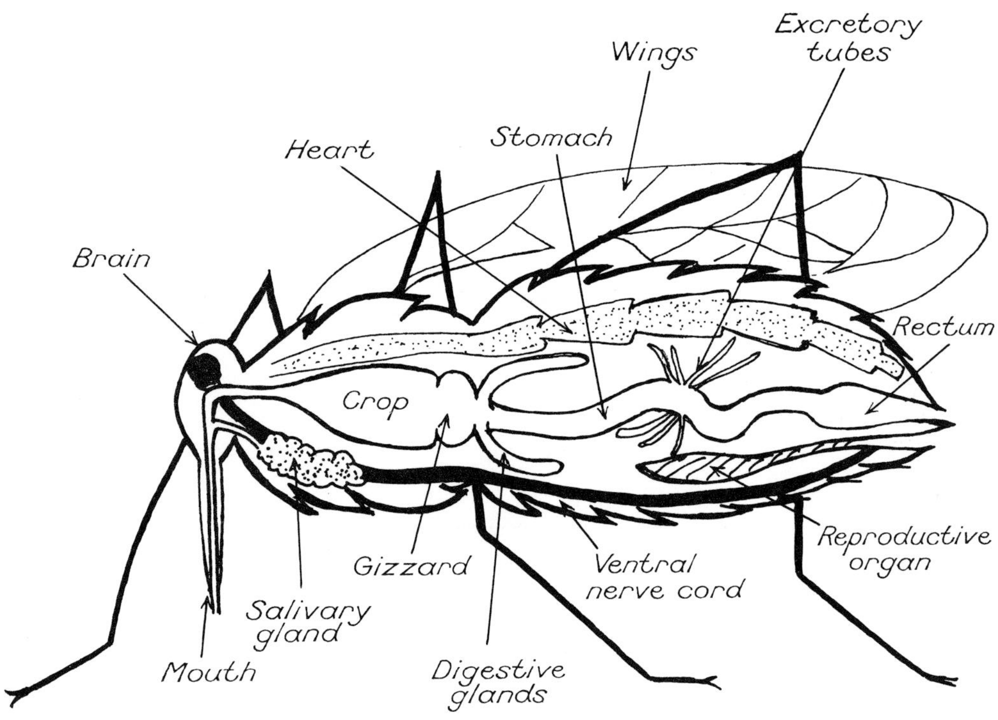

Fig. 3.4 General internal structure of an insect

Flies and disease

The following table shows the location of and the diseases transmitted by various flies:

Name	Location	Disease transmitted
Mosquito: Anopheles	Breeds in ponds, puddles of still water Africa, America, India, Australia	Malaria—protozoan disease
Mosquito: Culex	Breeds in pools, and puddles of still water in tropical countries (Fig. 3.5)	Filaria nematode worm causing elephantiasis
Tsetse fly (Plate 3, p. 19)	In soil, and in vegetable waste. Africa	Sleeping sickness—protozoa disease
Bluebottle or blowfly	Eggs laid in dead decaying matter	Similar diseases to that transmitted by housefly

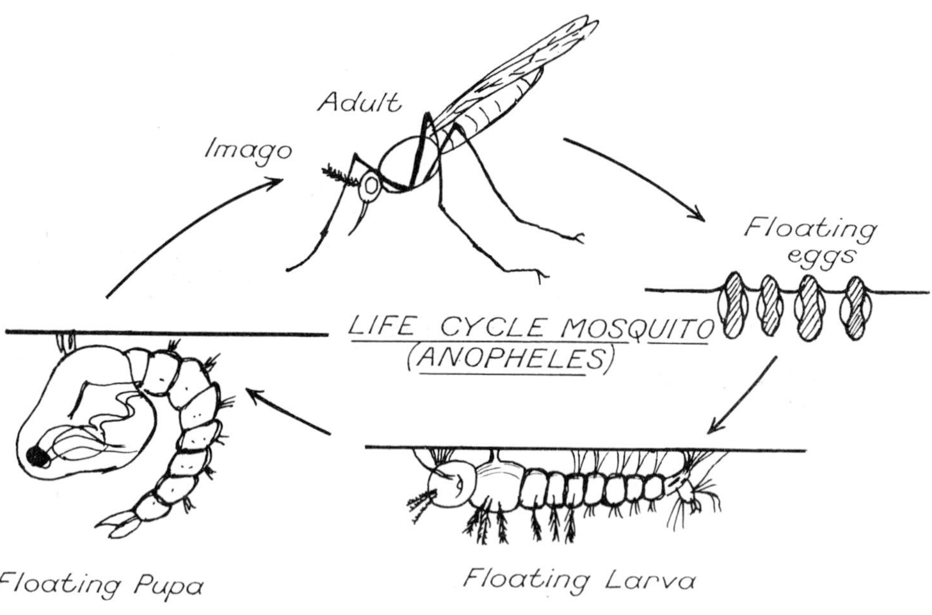

Fig. 3.5 Mosquito life cycle

LICE (ANOPLURA)

Sucking lice found as parasites on man include the body, head and crab or pubic louse.

General features

In addition to having some of the features seen in insects the following are typical of lice:

(a) Small grey coloured, *wingless* insects covered in hairs.
(b) Legs are adapted to clinging to the human skin or hair through the claws.
(c) Mouth parts are for piercing the skin and sucking blood.
(d) Eggs are cemented to hairs or clothing, to hatch into nymphs that become imagos in 3 weeks of hatching.

Human infestation

Lice are *ectoparasites* in contrast to *endoparasites* such as tapeworms. They cause itching and their bites are infected by microbes from their saliva; in addition the skin abrasions caused by scratching with unclean finger nails may become infected with saliva and faeces of the louse. Diseases transmitted by lice include *typhus fever* (a disease showing as a high temperature and rash) *trench fever* and *relapsing fever.*

Plate 6 The louse, an ectoparasite on the hair and skin of man (World Health Organisation)

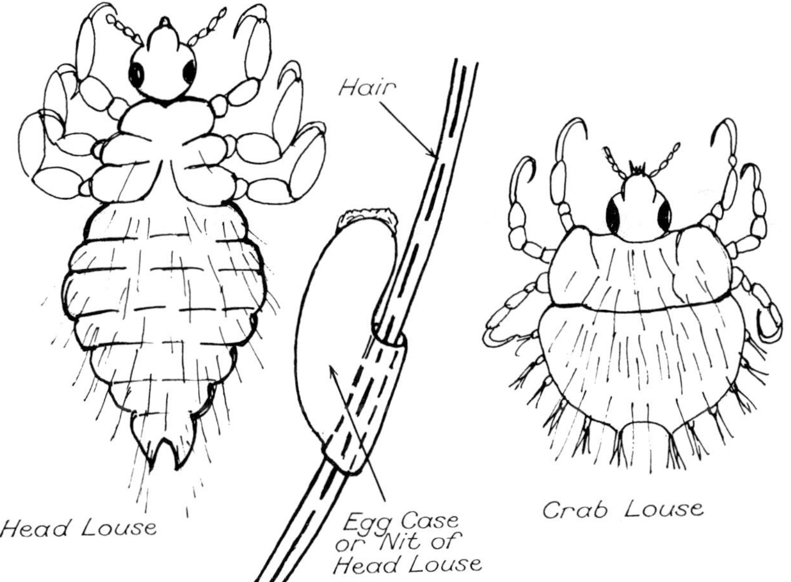

Fig. 3.6 Lice

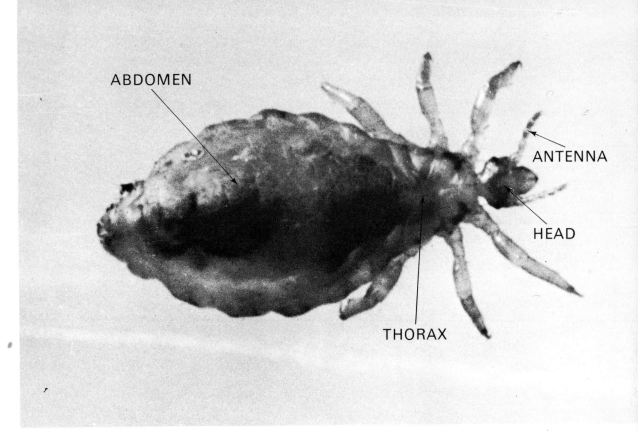

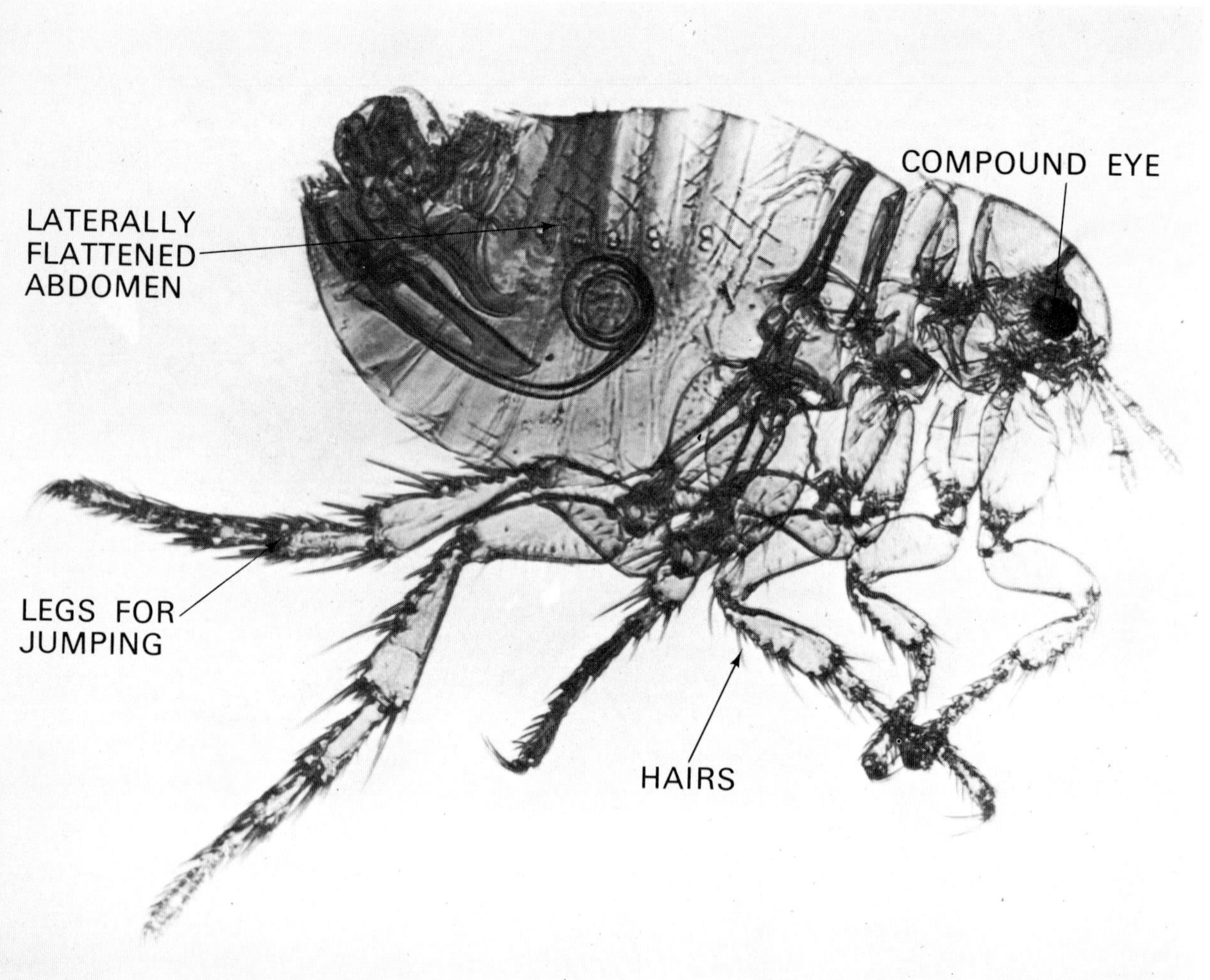

Plate 7 The human flea, an ectoparasite with biting and sucking mouthparts (Rentokil)

FLEAS

Fleas are typical *wingless* insects found as ectoparasites on hairy or furry animals and also on feathered poultry.

General features

(1) Small dark brown insects covered in hairs.
(2) Segmented body is flattened from side to side.
(3) Three pairs of legs are adapted for *jumping* allowing rapid transmission from host to host.
(4) *Biting* and *sucking* mouthparts pierce the skin injecting saliva to prevent clotting of blood; the saliva is infected with microbes which are transmitted by injection to the host.

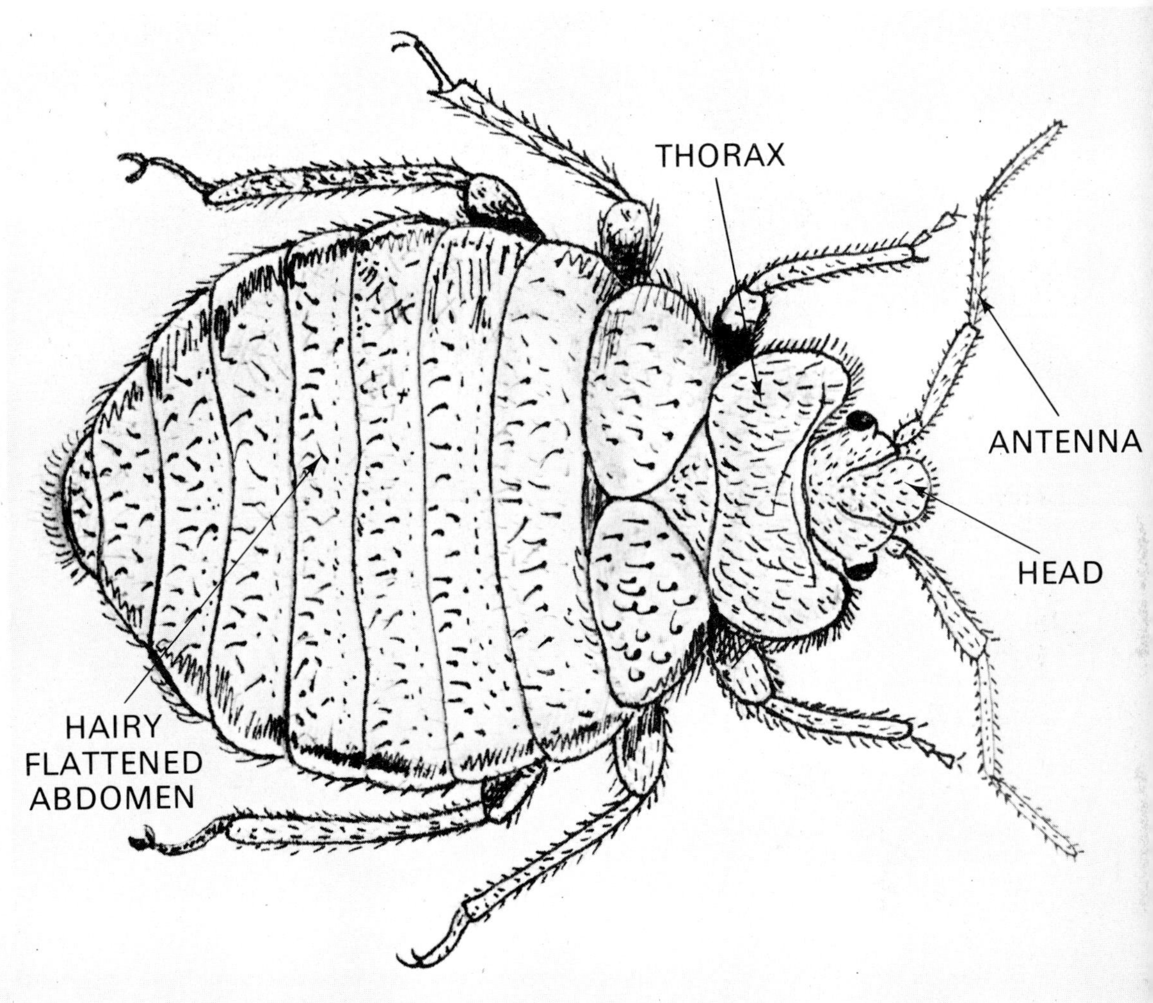

Plate 8 The bed bug (Rentokil)

(5) Irritation of the host skin by the flea's hairs results in scratching and introduction of flea faeces into abrasions and bites frequently through unclean finger nails.
(6) Flea *metamorphosis* is *complete*. The egg hatches in the filth of cracks and crevices of houses into a pupa, from which a flea imago emerges to seek a warmblooded host.

Human infestation

Rat fleas transmit *bubonic plague* in man. This is a highly infectious disease transmitted further by droplet infection from the coughing of an infected person.

Dog tapeworms are known to pass to man by accidental swallowing of dog fleas.

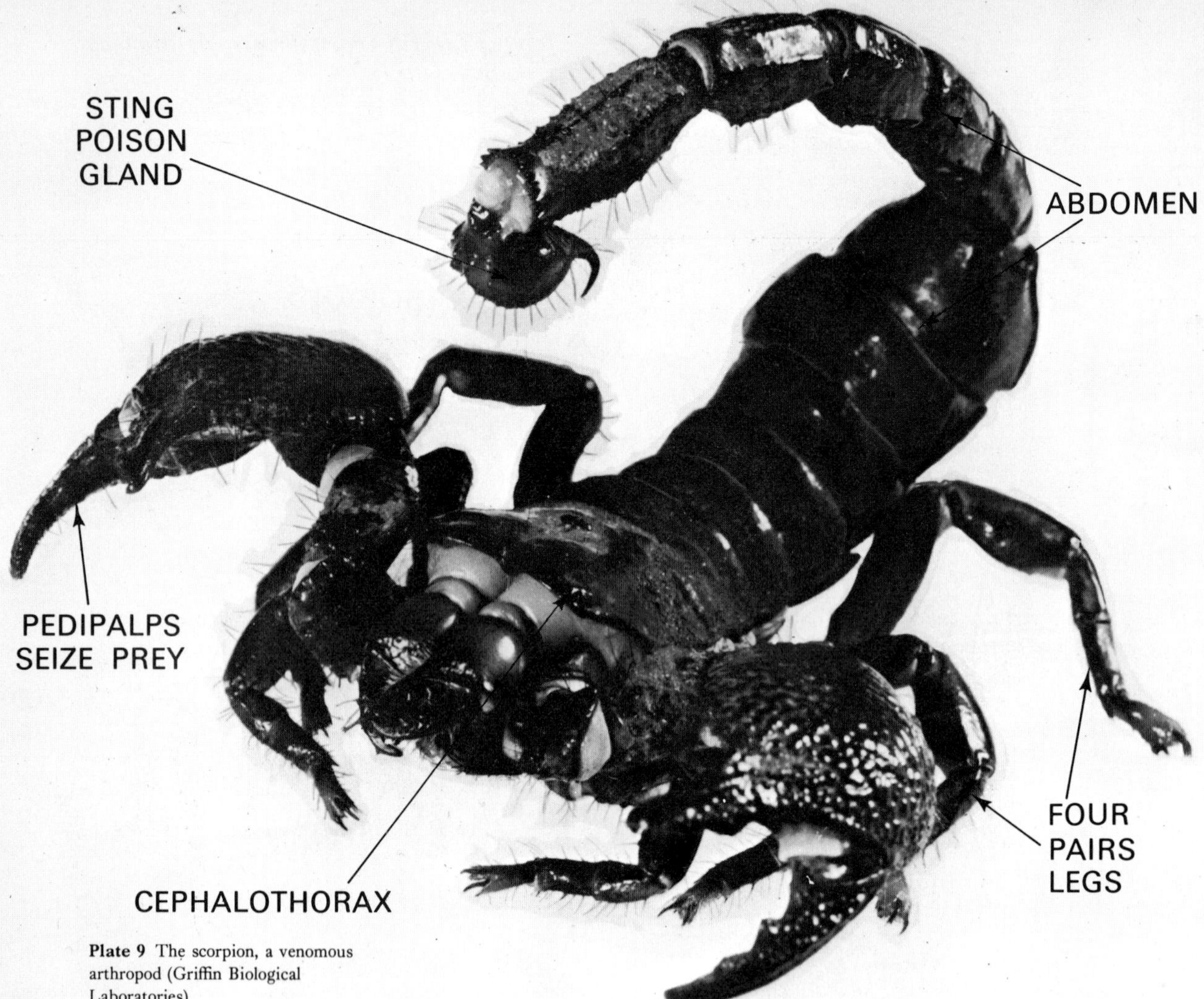

Plate 9 The scorpion, a venomous arthropod (Griffin Biological Laboratories)

BUGS

Bedbugs are *nocturnal* animals emerging from their hiding places in cracks and crevices in filthy rooms at night.

General features

(a) They are larger than fleas, have brown coloured segmented bodies flattened from top to bottom and are covered in large hairs.
(b) Three pairs of legs are suited for swift running movement in order to return quickly to their hiding places.
(c) Bedbug bites are persistently itchy and can become infected through saliva injection and scratching from the host.
(d) A *peculiar odour* is emitted by the bedbug which helps to disclose its presence.
(e) Eggs are laid by being cemented to mattresses, in wall cracks, or behind wallpaper. Secondhand bedding, and furniture may introduce infestation into a home.

The eggs hatch into nymphs, which after moulting become adult imagos by incomplete metamorphosis.

OTHER ARTHROPODS CAUSING DISEASE IN MAN

Spiders, scorpions, mites and ticks form a class of arthropods called *Arachnida* which have the following general features:

(a) The body is in *two* parts a *cephalothorax* or combined head and thorax, and a segmented *abdomen.*
(b) Four pairs of legs are present in contrast to the three pairs seen in insects.
(c) Mites and ticks have biting and sucking mouthparts.
(d) *Venom* or *poison* glands are found in certain spiders and scorpions.

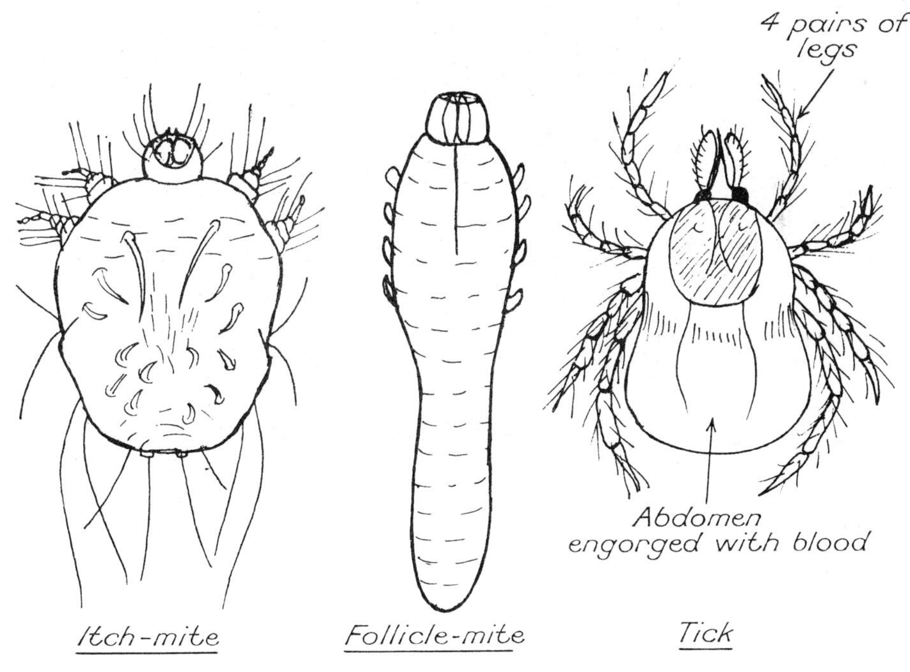

Fig. 3.7 Mites and ticks

MITES

These are tiny animals often seen only by means of a microscope.

Itch mite: scabies burrows into the skin, the female laying eggs at the end of tunnels in the skin found between the fingers and toes. Transmission is by direct close contact in holding hands or sleeping together. Symptoms are seen as a red thickened skin with infected spots and pencil line tunnelling.

Follicle mite is a wormlike mite found in the hair follicle or sebaceous gland of human skin, causing a burning, itching, red, scaling skin.

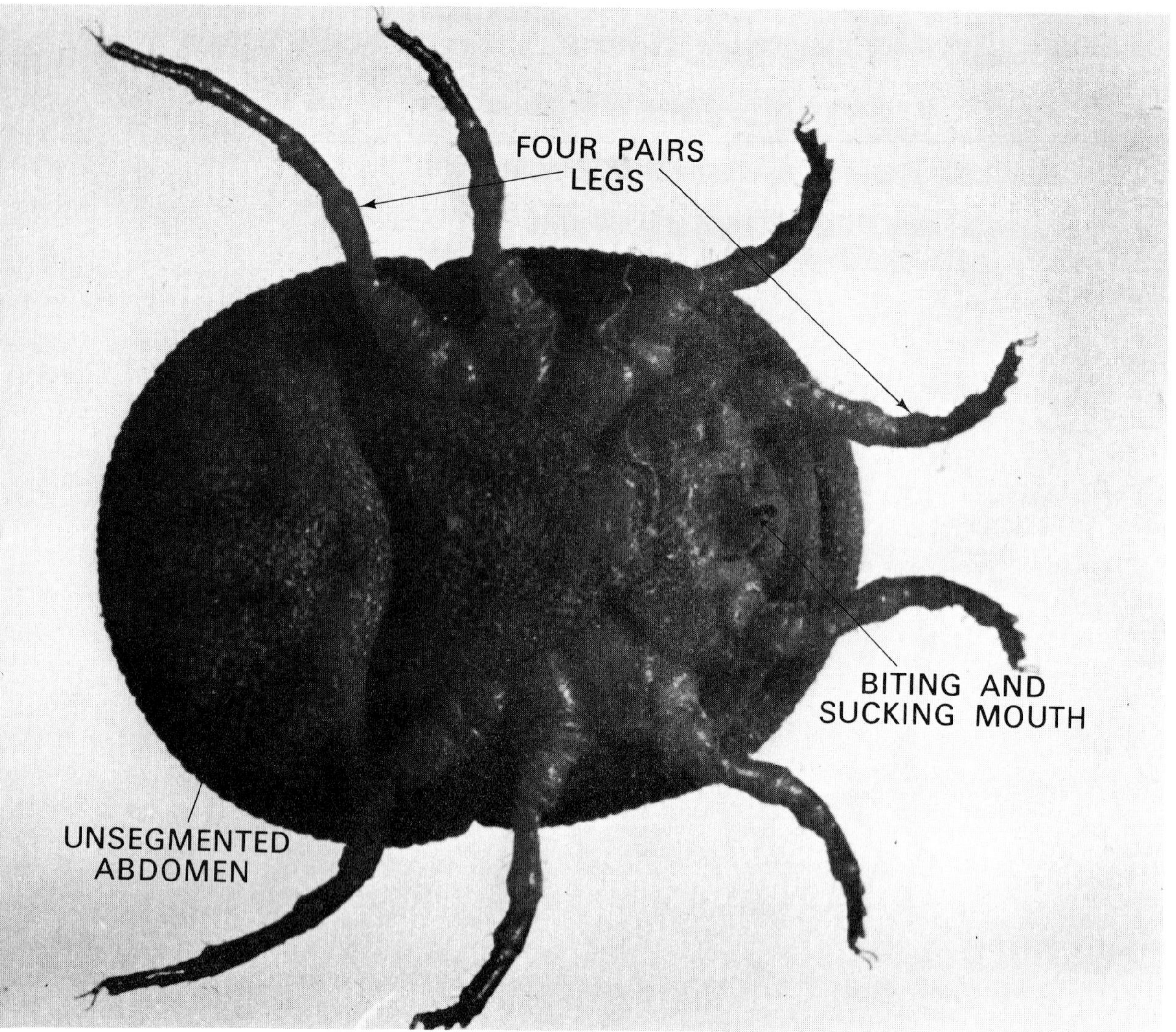

Plate 10 A tick, an arthropod responsible for transmitting relapsing fever and encephalitis (World Health Organisation)

TICKS

These are easily visible to the naked eye as oval shaped animals with four pairs of legs. The tick attaches itself to the skin of the host by special hooked *chelicerae* which make them very difficult to pull away from the skin. Their mouthparts cut and pierce the skin in order to suck blood, their bodies become engorged with blood and the tough leathery body swells and falls to the ground.

Eggs are laid in the soil and hatch into nymphs that attach themselves to the legs of passing animals, climbing upwards and attaching themselves to the abdomen by chelicerae. Ticks transmit *tick fever*, *encephalitis*, and *typhus fever*.

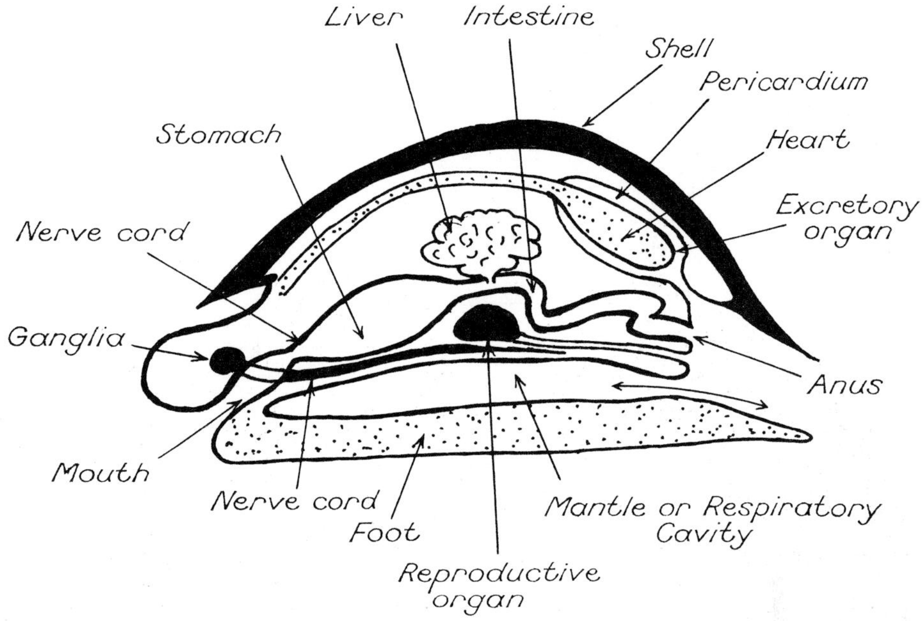

Fig. 3.8 General structure of a mollusc

SHELLFISH: MOLLUSCA

Cockles, mussels, snails, and octopuses are a group of animals that are members of the phylum mollusca. Their main features apart from possessing a *shell* in some form or another are summarised as follows:

(1) One or two *shells* are found either *outside* the body as in oysters, or embedded *within* the body as in cuttlefish. Snails have a single shell whilst cockles have two shells.

(2) A complex *digestive system* is found compared to most other invertebrates. Simple *teeth* are needed to cut up the food, which is also digested by secretions from digestive glands similar in function to the mammal liver and pancreas. A long coiled intestine serves to allow the food to take longer to travel along its length allowing more complete digestion and absorption of the food.

(3) A *heart* composed of an *auricle* and a muscular *ventricle* contained within a pericardium, shows the beginnings of a chambered heart which is found developed to a greater extent in the vertebrates.

(4) *Lung* spaces or a mantle cavity seen in snails, or filamentous *gills* are seen in mussels; both are well supplied with blood vessels and have the most highly developed respiratory systems found in invertebrates.

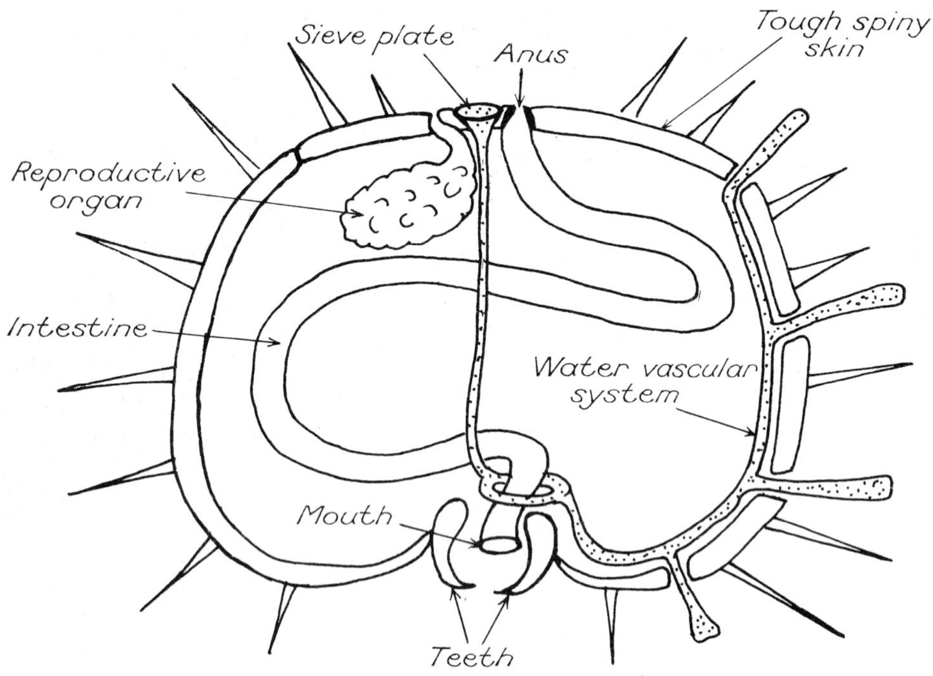

Fig. 3.9 General structure of a sea anemone or echinoderm

Molluscs and man

Freshwater snails are the intermediate hosts in transmitting liver fluke diseases in sheep and other grazing animals.

Cockles and mussels can become contaminated with microbes present in sewage, causing *typhoid* and similar diseases; strict cleansing of gathered shellfish is always carried out in special cleansing tanks. Certain shellfish have in the past become infected with a green algae, *Gonyaulux*, which makes the shellfish highly poisonous, causing symptoms of the disease seen as muscular paralysis, thirst, dizziness, and frequently the disease is fatal.

SPINY SKINNED STARFISH: ECHINODERMS

Starfish, sea-urchins, and brittle stars, are marine invertebrates with typically spined and prickly skins. Starfish, although having little if any direct effect on man, are included in this text to present a complete picture of the outline survey of animals which are invertebrates.

General features

(1) *Radial symmetry* is seen in the body form, where the starfish shows a radiating structure like the spokes of a wheel. Most invertebrate animals show *bilateral symmetry* or an arrangement into right and left sides, their bodies can only be halved in one plane to produce two parts which are mirror images of each other.

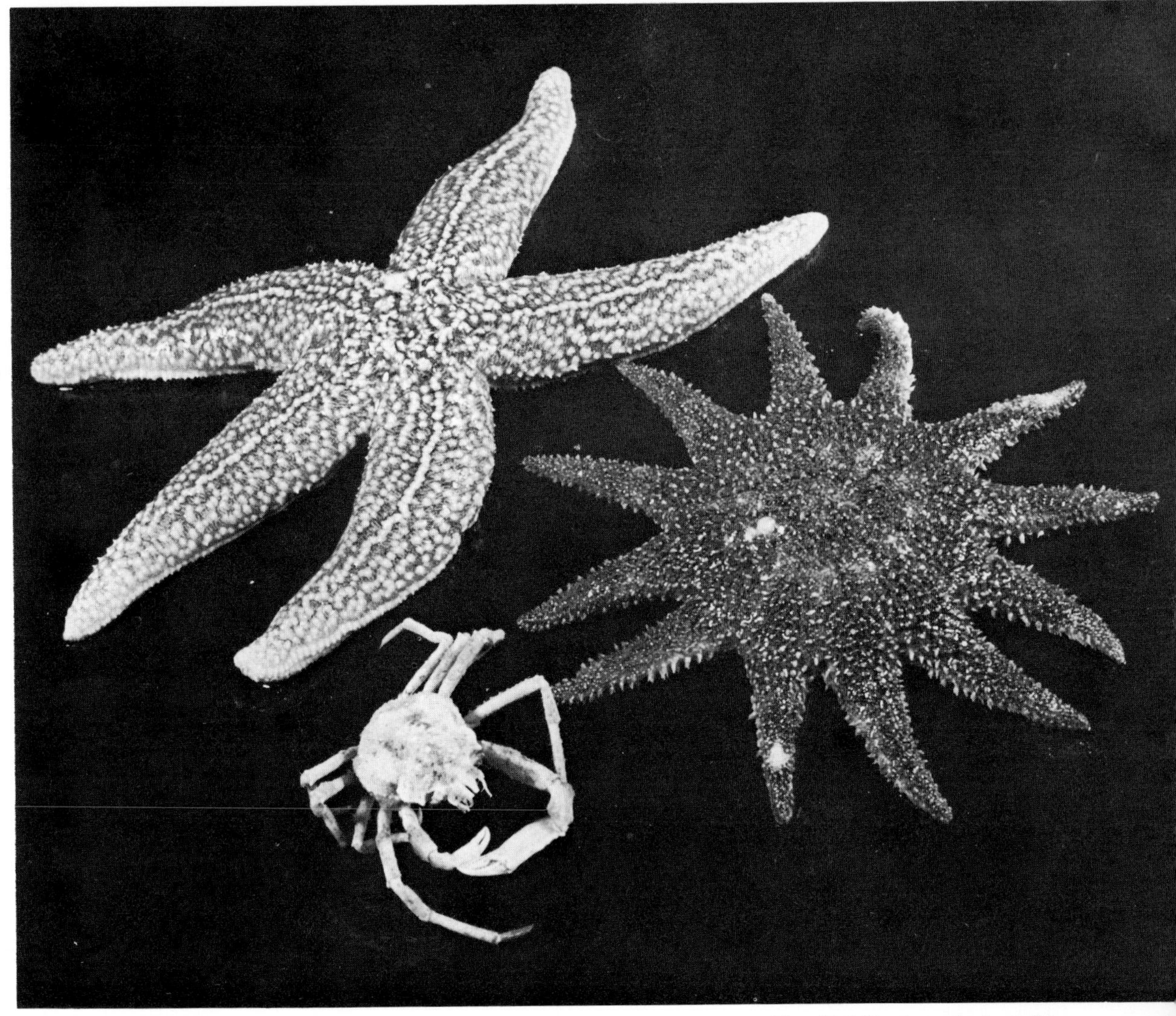

Plate 11 Echinoderm animals, starfish and sunstar, with an arthropod stone crab, all showing exoskeletons (Ministry of Agriculture, Fisheries and Food)

(2) The *exoskeleton* consists of spines and plates serving to protect the soft internal organs.
(3) Feeding is by means of an *extensible* stomach which surrounds the food.
(4) The *water vascular system* which employs seawater for the purpose of moving the animal through tube feet and also serves to carry oxygen for respiration and act as a substitute for blood fluid. A blood system of blood cells or fluid is not present.

4 *Vertebrate Animals*

Vertebrate animals are those which possess a spine or vertebral column.

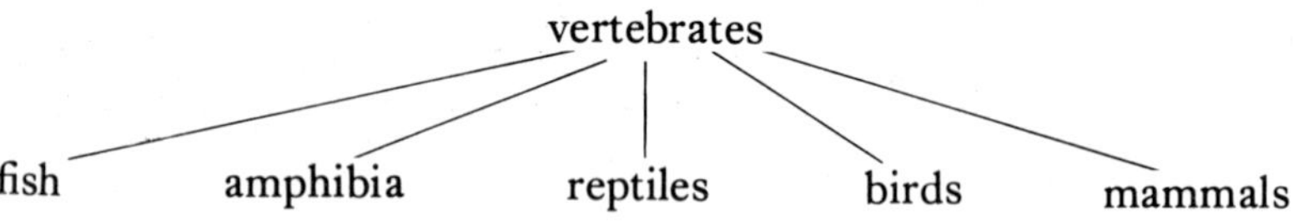

FISH AS VERTEBRATES

BODY STRUCTURE

To live in water it is necessary for fish to be specially adapted.

Water is a resistant medium, causing the body of a fish to become streamlined and tapering in shape. The body consists of a *head*, *trunk*, and *tail*, without any neck. Fins form *limbs* as paired structures which manoeuvre the fish through the water propelled by the tail.

Fish are multicellular, coelomate, animals in which the cells are arranged as *ectoderm*, *endoderm* and *mesoderm*. A transverse section of a fish is shown in the figure of a cod steak; the important structural features of a vertebrate animal are seen in this sectional view.

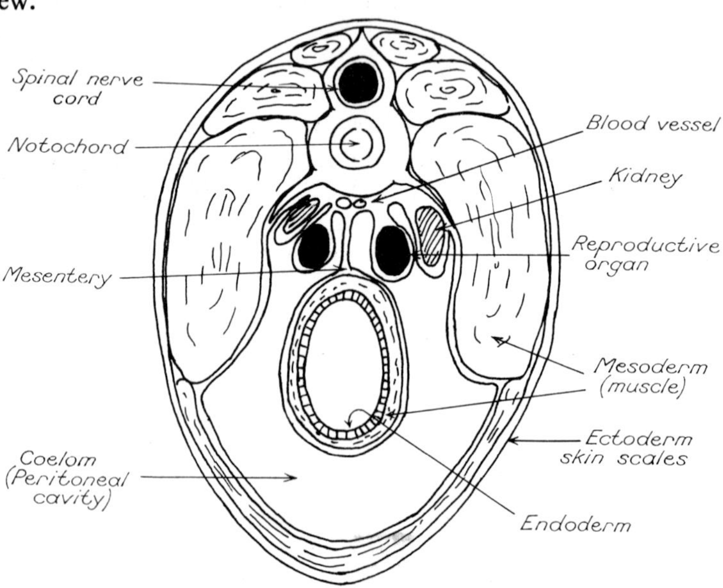

Fig. 4.1 Sectional view of a fish

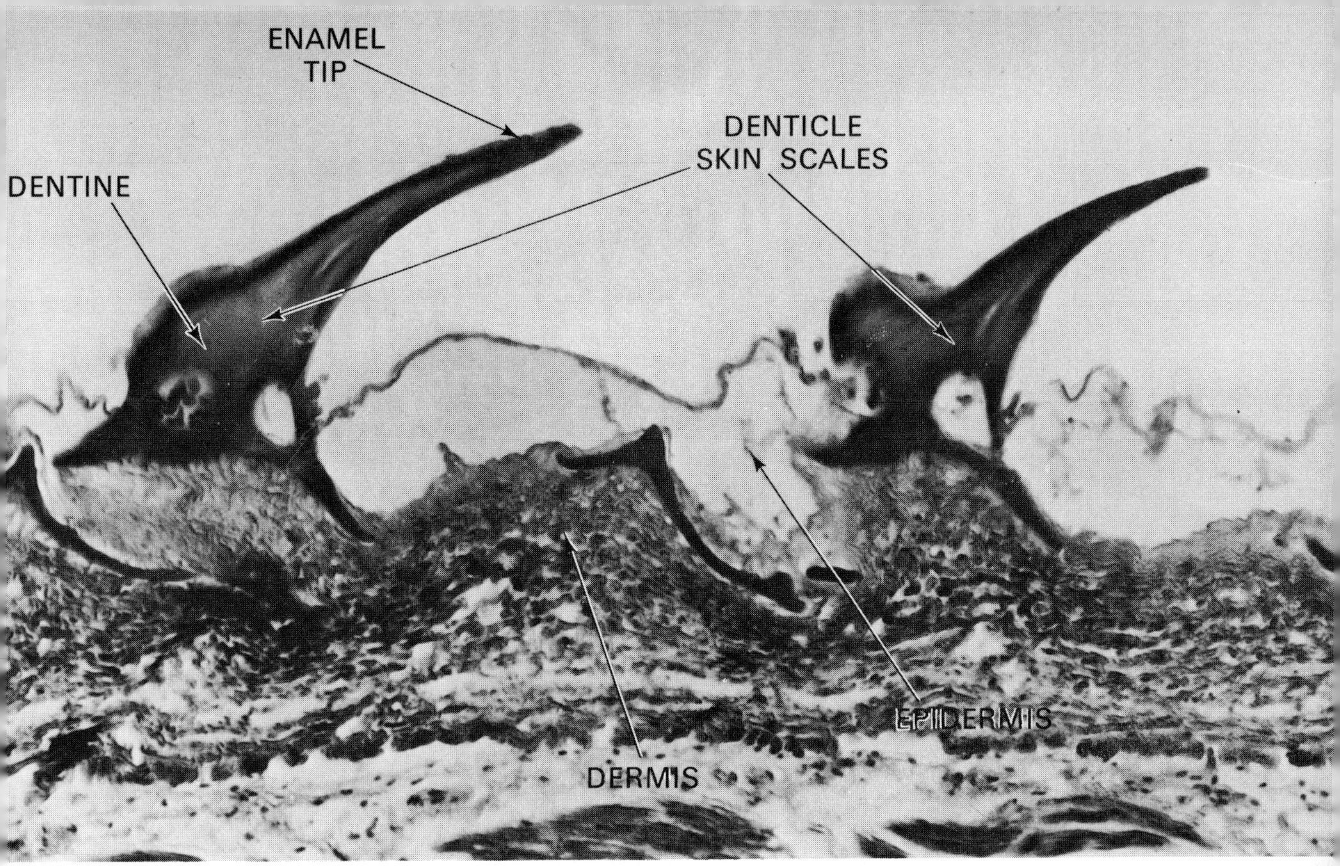

Plate 12 Skin of the dogfish showing toothlike denticles or skin scales (Griffin Biological Laboratories)

(1) **Skin.** This is composed of *scales* kept moist by slimy *mucus.* In sharks and dogfish the skin scales resemble miniature teeth.

(2) **Muscle.** Fillets or long slices of fish, show the arrangement of the body wall muscles in blocks or *segments* called *myotomes*, this feature indicates the segmental structure of a vertebrate body.

(3) **Gut.** The *alimentary canal* or gut is lined with *endoderm* and the muscular wall forms from the *mesoderm.* Around the gut is the abdominal or *peritoneal* cavity which is a space formed from the *coelom.* A similar coelomic space surrounds the heart in the *pericardium.* The gut is supported by *mesenteries* within the peritoneal cavity.

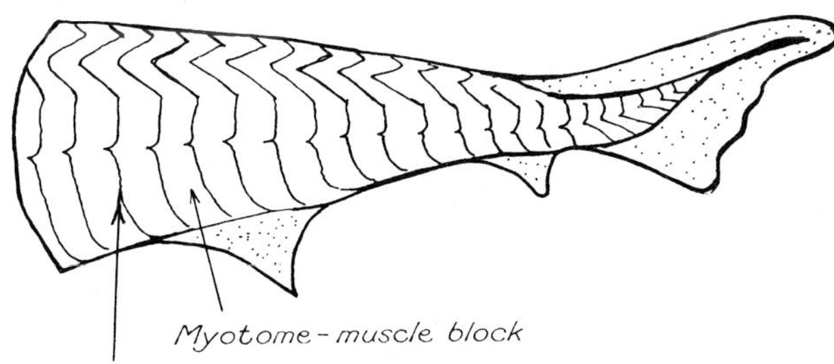

Fig. 4.2 Segmentation of muscle in fish

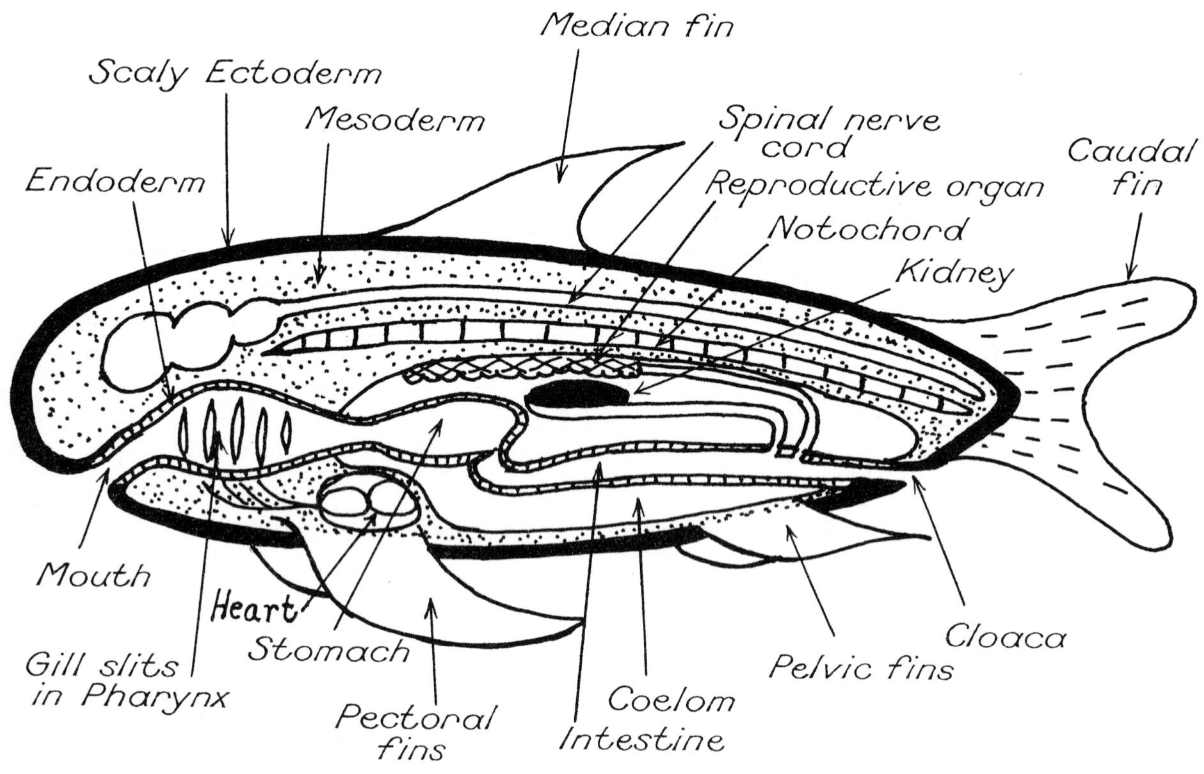

Fig. 4.3 General structure of a fish

(4) **Skeleton.** Vertebrates have *endoskeletons*, the first formed part being the *notochord rod* which is later surrounded by bone forming individual *vertebrae*, ligaments join them together in a vertebral column. Protecting the nerve cord is the bony *neural arch*, whilst the notochord becomes the *centrum* of a vertebra.

Skeleton bone can be either *cartilage* or *hard bone*. Dogfish and sharks have soft skeletons of cartilage, whilst other fish, whiting and herring, have a hard bony skeleton in which the cartilage has become *ossified* by calcium and phosphorus salts.

NUTRITION

Plankton forms the food of some fish. It is drawn into the mouth where the water is passed out through *gill* slits in the *pharynx*, here gill rakers remove the plankton. Other fish such as sharks have separate upper and lower jaws which help to *hold* the food, but not chew it. The upper jaw is loose and not bound to the cranium as in man.

Complex foods are digested by means of *enzymes* produced partly by the endoderm lining the stomach, duodenum, intestine and liver. Absorption of the digested food is by way of the endoderm lining the intestine.

A *cloaca* serves as the common opening for the discharge of faeces, urine, and reproductive products.

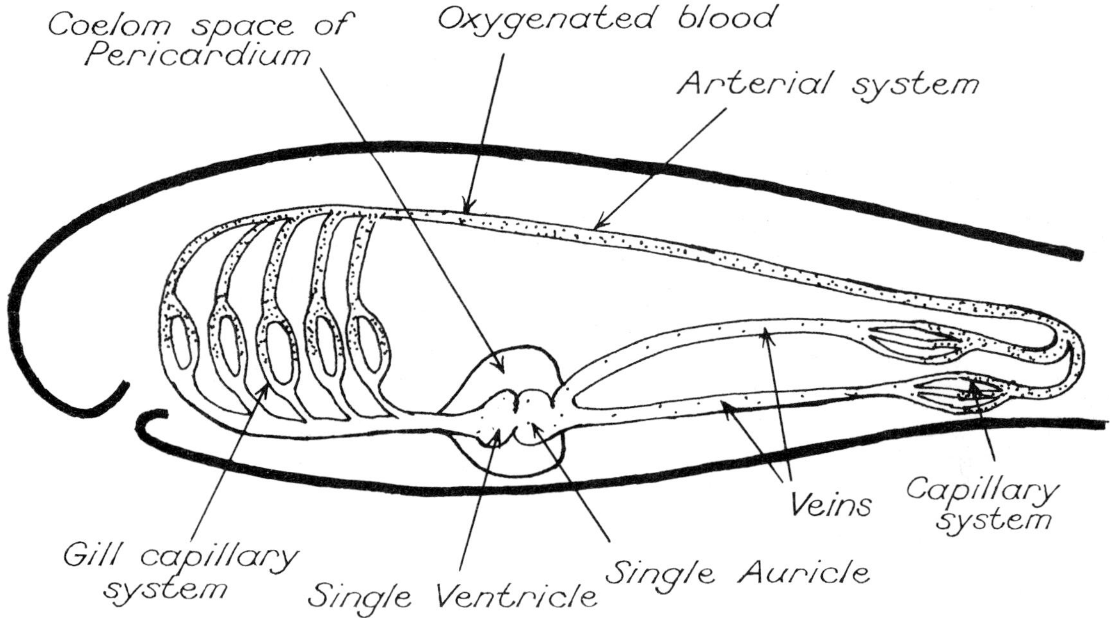

Fig. 4.4 Blood circulation in a fish

CIRCULATORY SYSTEM

The blood system consists of fluid *plasma* closely resembling sea water, and blood *cells* containing the red oxygen carrying pigment *haemoglobin*, this is circulated through the body by the muscular *heart*, located in a ventral position. The heart is two chambered, consisting of one *auricle* or *atrium* and one *ventricle*. Blood vessels or *veins* carry blood to the gills to receive oxygen, the oxygenated blood being distributed to the body cells by *arteries*. Deoxygenated blood collects in large blood spaces called *blood sinuses* connected to vessels which return the blood into the heart. The heart contains only deoxygenated blood as shown in Fig. 4.4.

RESPIRATION

Oxygen is more soluble in freshwater than in seawater; the amount of oxygen dissolved in freshwater is thirty times *less* than the oxygen present in air. Consequently large amounts of water are drawn over the gills of fish to obtain the necessary oxygen. The dissolved oxygen *diffuses* rapidly through the moist, thin membranes of the gill *capillary* blood vessels whose walls are only one cell in thickness (Fig. 4.5).

Fish, amphibia and reptiles are cold-blooded animals and are called *poikilothermic*; this means their body temperature varies and is slightly higher than the temperature of their surroundings.

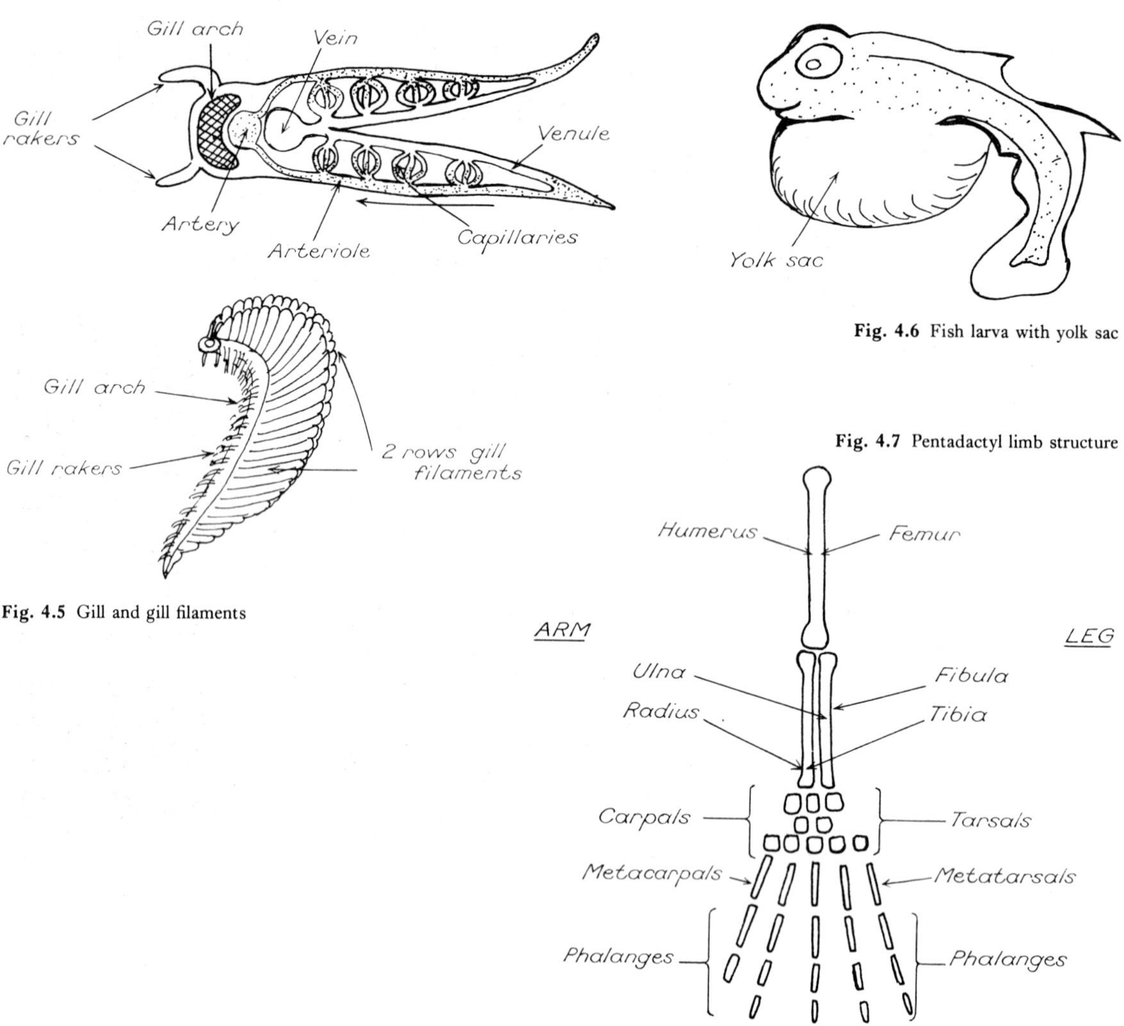

Fig. 4.5 Gill and gill filaments

Fig. 4.6 Fish larva with yolk sac

Fig. 4.7 Pentadactyl limb structure

EXCRETION

Kidneys exist for the purpose of filtering waste from the blood and also to remove the large amount of water which enters the body. Urinary bladders are not found in fish, the urinary waste being discharged frequently through the cloaca.

NERVOUS SYSTEM

A *nerve cord* is found in a *dorsal* position and forms from the *ectoderm* layer in all vertebrates; this contrasts with ventral position of the nerve cord seen in invertebrates (Fig. 4.3).

The brain develops at the anterior end, this position being the one which encounters stimuli head on. Ten *cranial* nerves arise in pairs from the brain compared to twelve pairs found in man.

Fish have a keen sense of sight, smell and balance. The ear of a fish is primarily an organ of balance consisting mainly of *semicircular* canals.

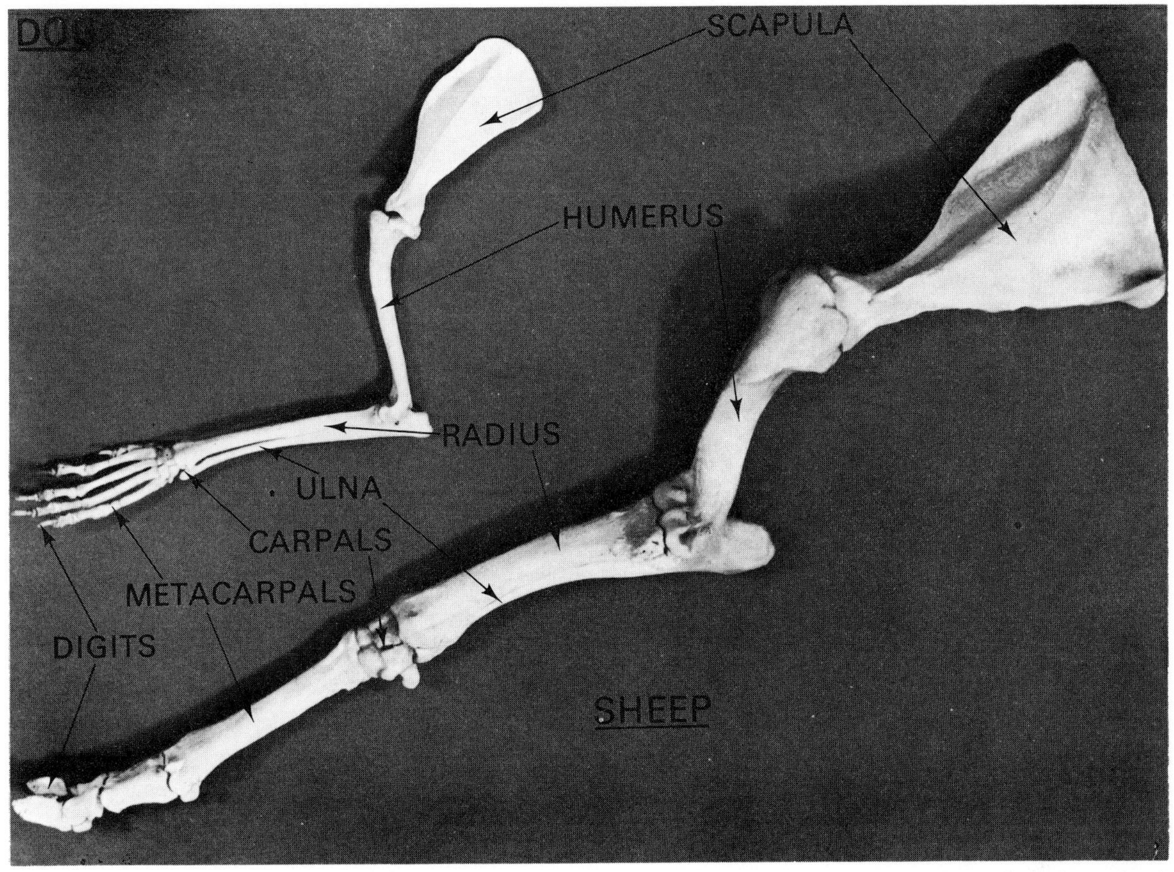

Plate 13 Bones of typical pentadactyl limbs of a dog and sheep (Griffin Biological Laboratories)

REPRODUCTION

Vertebrates show distinct sexes of male and female. Great numbers of eggs are laid by the female fish which are usually fertilised outside the body by *external fertilisation*; seldom are the eggs fertilised internally.

Each egg has a small amount of *yolk* to feed the developing embryo which is completely surrounded by the *watery environment* of the sea, freshwater or in the egg case, this environment being essential for the development of vertebrate embryos (Fig. 4.6).

AMPHIBIA AS VERTEBRATES

Amphibians are those vertebrates which have the ability to live partly on land and partly in water: such animals include the frog, toad, and newt.

BODY STRUCTURE

Amphibians, being multicellular, three layered, coelomate, animals, have a skin produced from the ectoderm which is very thin and kept continually moist by mucus.

The limbs of land-living vertebrates have a typical structure of five fingers or toes, called *pentadactyl* limbs. See Fig. 4.7 and Plate 13.

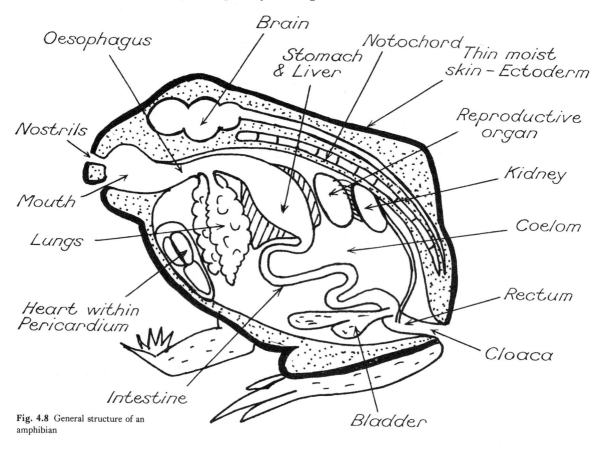

Fig. 4.8 General structure of an amphibian

NUTRITION

Teeth in the form of simple pegs serve to hold but not to chew the food. The digestive system follows the general pattern of that found in vertebrates consisting of a stomach, intestine, and rectum. See Fig. 4.8.

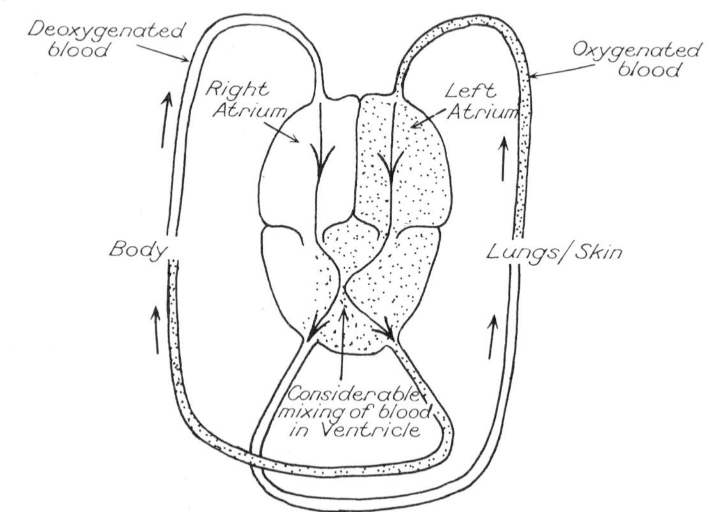

Fig. 4.9 Blood circulation in an amphibian heart

CIRCULATORY SYSTEM

Blood is pumped by the heart through a system of blood vessels, consisting of *arteries* and *arterioles*, linked by the *capillary vessels* to *venules* and *veins*.

The heart of an amphibian consists of *two auricles* or atria, and *one ventricle*. The blood flow is shown in Fig. 4.9. Oxygenated blood from the lungs and *skin* is pumped by the ventricle to supply body cells with food and oxygen. The deoxygenated is received by the *left* auricle, whilst the oxygenated blood is received by the *right* auricle. It is important to note the *partial mixing* of oxygenated and deoxygenated bloods which occurs in the ventricle.

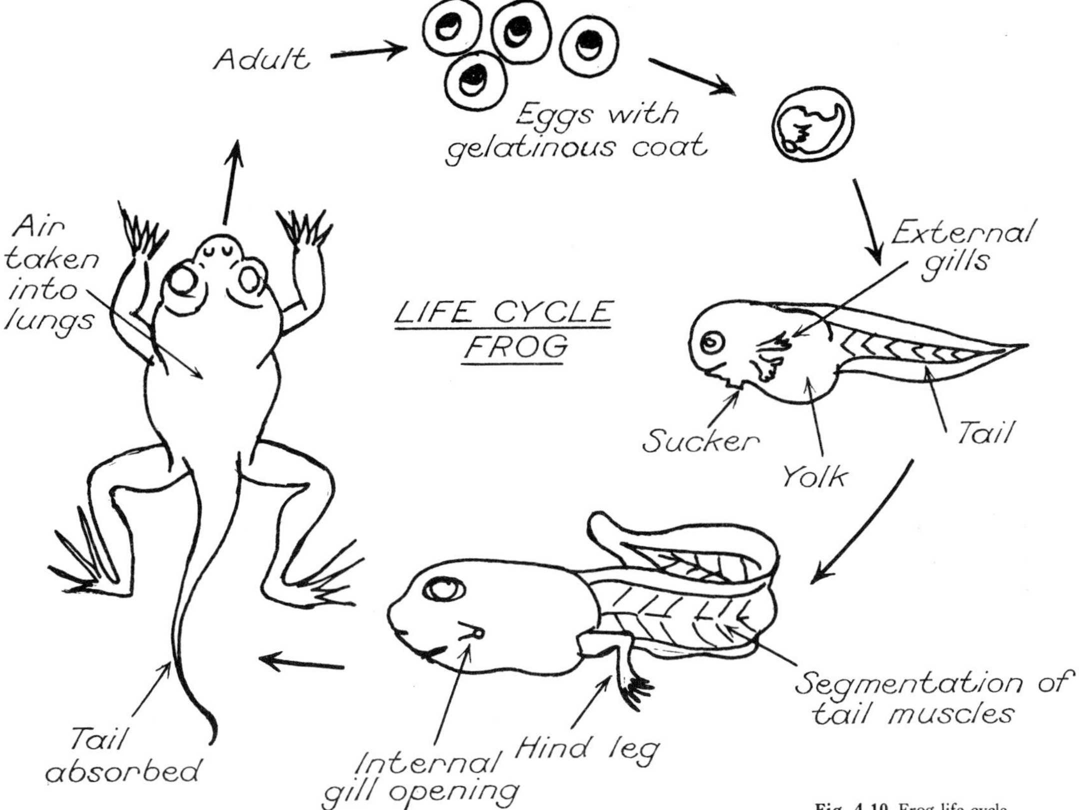

Fig. 4.10 Frog life cycle

REPRODUCTION

A number of eggs are laid by the female amphibian which are fertilised *externally* by the male pouring spermatozoa onto them as they enter the water. The fertilised eggs pass through a process of change or *metamorphosis*, which is summarised in Fig. 4.10. The watery environment provides fairly steady environmental conditions in a steady temperature, and supply of oxygen, together with a rapid means of disposing of excretory waste into the surrounding water in which it is rapidly diluted.

The developing young amphibians are not *cared for* by the parents and consequently the mortality rate amongst them is high.

RESPIRATION

During the life cycle of amphibia the process of respiration may take place through *gills* in the tadpole *larva*, or *lungs* and the *skin* in the adult. Two thin-walled sacs, well supplied with blood vessels and continually moist, act as lungs. The frog's skin also offers a large *surface area* which is continually moist and is very thin, allowing oxygen of the air to *diffuse* rapidly into the blood within the capillary vessels.

The lungs are contained within the trunk and air is first drawn into the mouth where it is held for a short time, allowing some oxygen to diffuse into the *epithelium* lining the mouth. Air is forced down into the lungs by *positive pressure* when the mouth floor is raised and the nostrils kept closed. After entry and holding in the lungs, air is expelled by the pressure exerted on the lung sacs by the contents of the trunk, and contraction of body wall muscles.

NERVOUS SYSTEM

Amphibia have developed a sense of *hearing*, by having ears which are made up of a *tympanic membrane* or eardrum connected to a *middle ear* by a single bone or *ossicle*. The inner ear is concerned with balance through the semicircular canals, and in sound detection through the *cochlea*. The *eustachian tube*, linking the mouth and middle ear, develops from a gill slit seen in the embryo tadpole.

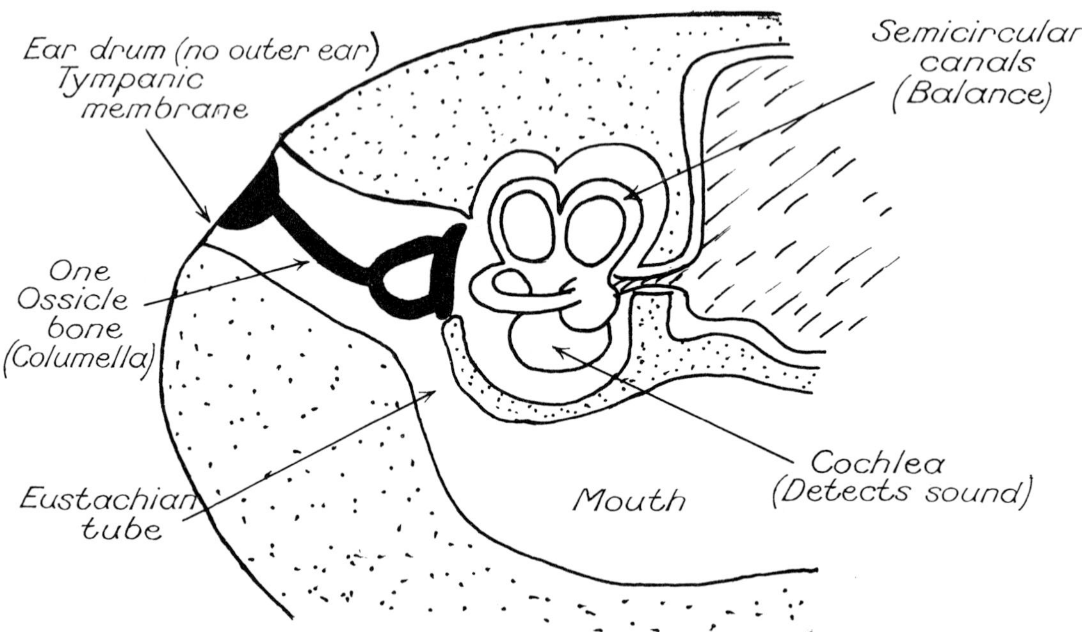

Fig. 4.11 Ear of an amphibian

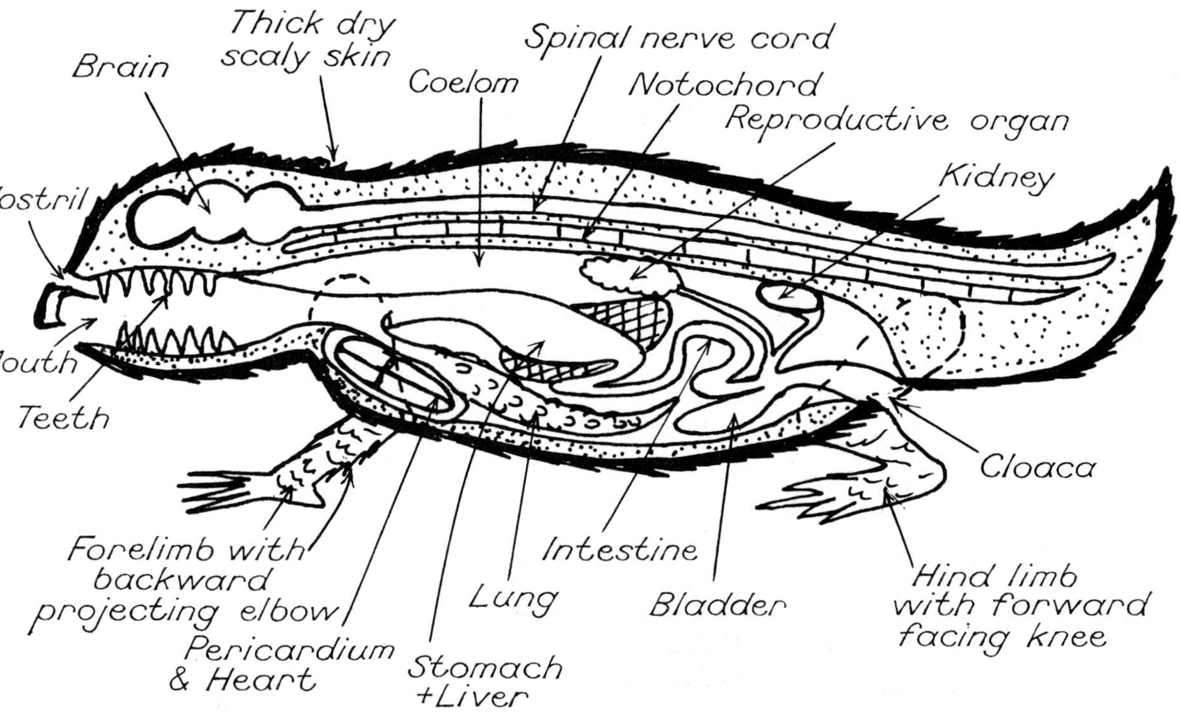

Fig. 4.12 General structure of a reptile

REPTILES AS VERTEBRATES

Reptiles are animals which include snakes, lizards, and tortoises. Reptiles were the first vertebrates to live their lives completely on dry land and show many adapted structural and functional features for this mode of life. Vertebrates living on dry land must contend with environment problems seen in changing temperatures, water shortage, and problems of disposing of excretory waste.

BODY STRUCTURE

The reptile body shows a division into *head*, *trunk* and *tail*, with a distinct *neck* separating the head and trunk. Land animals must support their bodies on typical *pentadactyl* limbs since they, unlike water living animals, cannot rely on water *buoyancy* to support them. Limbs are connected to the *appendicular girdles* by strong muscles.

The skin is covered in overlapping *scales* which serve to reduce water loss by *evaporation* through the skin. Some of the scales are thick and horny serve as a protective *exoskeleton* seen in the tortoise shell.

NUTRITION

There is a greater development of teeth and jaws in reptiles compared to amphibia. The teeth simply hold the food and prey and are not for chewing purposes. The alimentary system closely follows the normal vertebrate pattern as shown in Fig. 4.12.

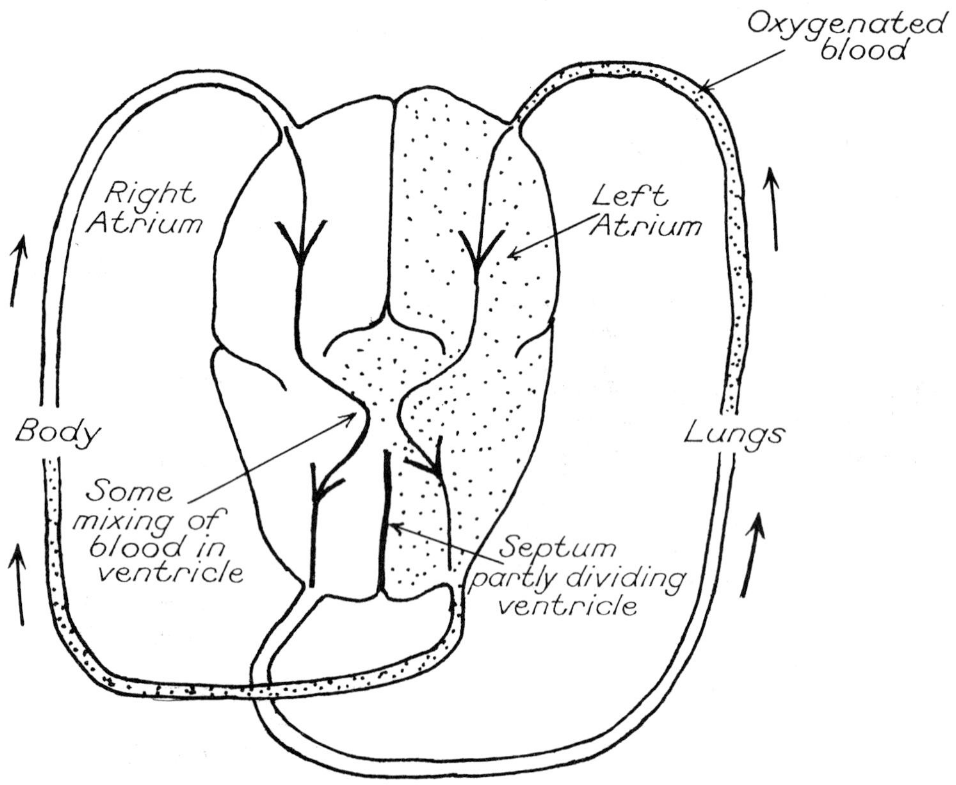

Fig. 4.13 Blood circulation in a reptile heart

CIRCULATORY SYSTEM

The heart is contained within a *pericardium*, and it consists of two chambers of the *auricles* or atria, with a single ventricle *partly divided* by a *septum*. The blood circulation is shown in Fig. 4.13, the right auricle receives the *oxygenated* blood, whilst the left auricle receives the *deoxygenated* blood. At some time there is a *mixing* of both deoxygenated and oxygenated bloods, but to a lesser extent than occurs in the amphibian heart. Usually the 'mixed' blood is circulated to the tail end of the body.

RESPIRATION

The thick, dry, scaly, skin prevents diffusion of oxygen through it, consequently the method of respiration is by means of *lungs*. Reptile lungs have their inner surface area increased by many tiny air sacs, or *alveoli* ,which are richly supplied by blood capillaries.

Air is drawn into the lungs by *relaxation* of the trunk wall muscles, and is forced out of the lungs by compression of the trunk wall by *contraction* of the trunk wall muscles.

EXCRETION

Reptiles excrete a concentrated form of urine from which most of the water has been removed in order to *conserve* the body water supply. The semi-solid urine, after a very short period of storage in the bladder, is eliminated through the cloaca with the faeces.

SENSE ORGANS

Reptiles have a keen sense of *sight*, whilst the ear shows the rudiments of an *outer ear* in addition to the inner and middle ears.

REPRODUCTION

Reptiles face the problem of water shortage by the following adaptions:

(a) fertilisation of the egg is *internal*, spermatozoa being introduced by copulatory organs.
(b) fertilised eggs are surrounded by *shells* which enclose the developing *embryo* within a miniature 'pond' of watery *egg white*, whilst food is provided in the *egg yolk*.

The eggs laid in soil or sand are *incubated* by the sun's warmth. Little if any *parental care* is shown in the hatching eggs with a consequent high mortality rate amongst the young.

BIRDS AS VERTEBRATES

Birds are the vertebrates mainly adapted to *flight* in the air, apart from *flightless* birds the kiwi and ostrich, whilst the penguin is an excellent *swimming* bird.

BODY STRUCTURE

The bird's body is organised into a *head*, with a distinct *neck* connecting to the *trunk* which has a short *tail*. Limbs of the leg hold the body in an *upright* position, whilst the forelimbs are modified into *wings* with three digits.

Feathers cover the skin, except on the legs which are covered with scales similar to reptile skin scales.

The *endoskeleton* consists mainly of *hollow* bones to reduce the weight of the body. Strong, well developed breast muscles are attached to the *keel* bone, an extension of the sternum, move the wings in flight. The limbs show the typical pentadactyl structure.

NUTRITION

Beaks are extensions of the upper and lower *jaws* and are without teeth. Food passes into the *crop* where it is softened before passing into the gizzard with thick muscular walls which together with a few purposely swallowed stones aid grinding of the food. The intestine terminates in a cloaca.

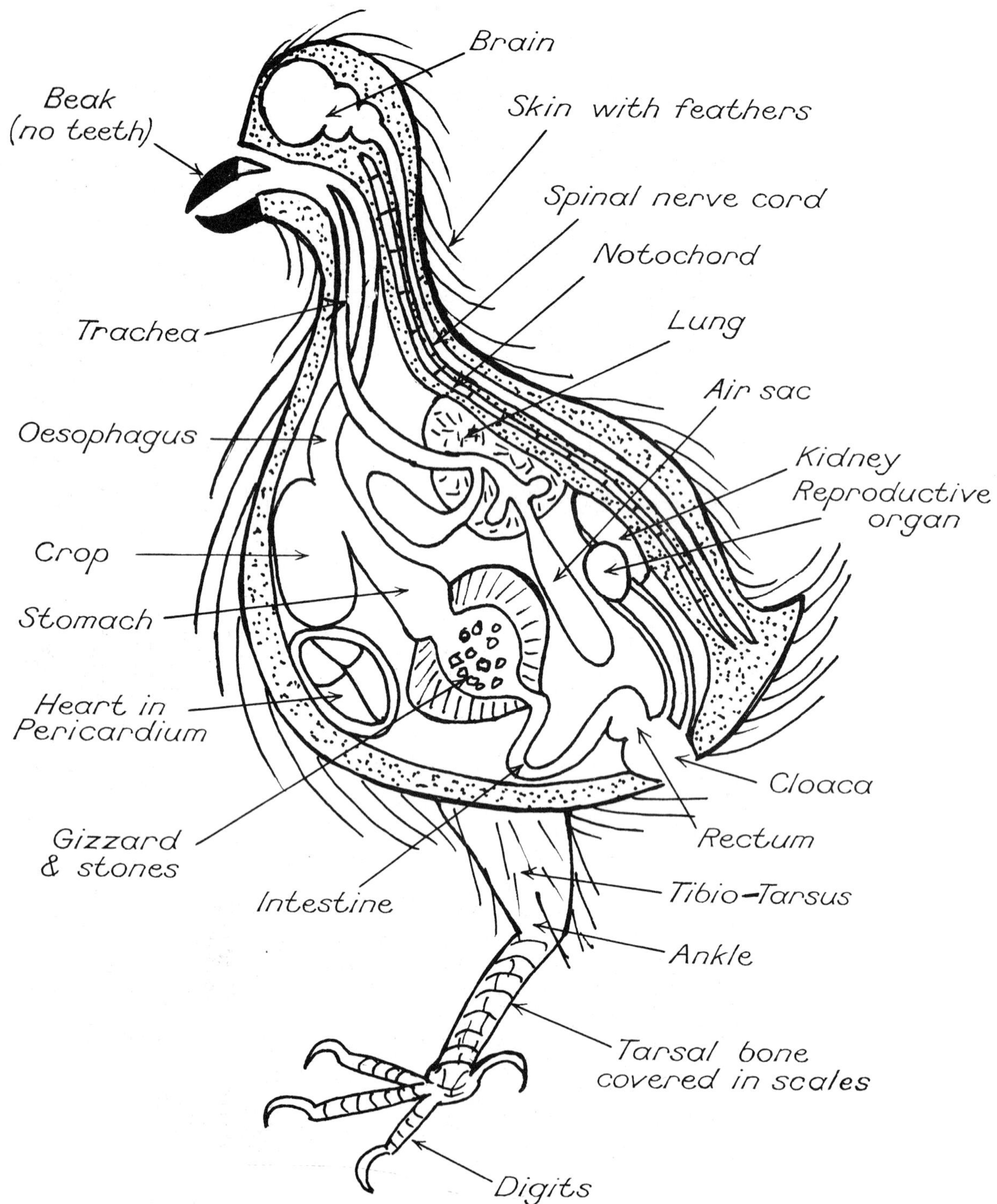

Fig. 4.15 General structure of a bird

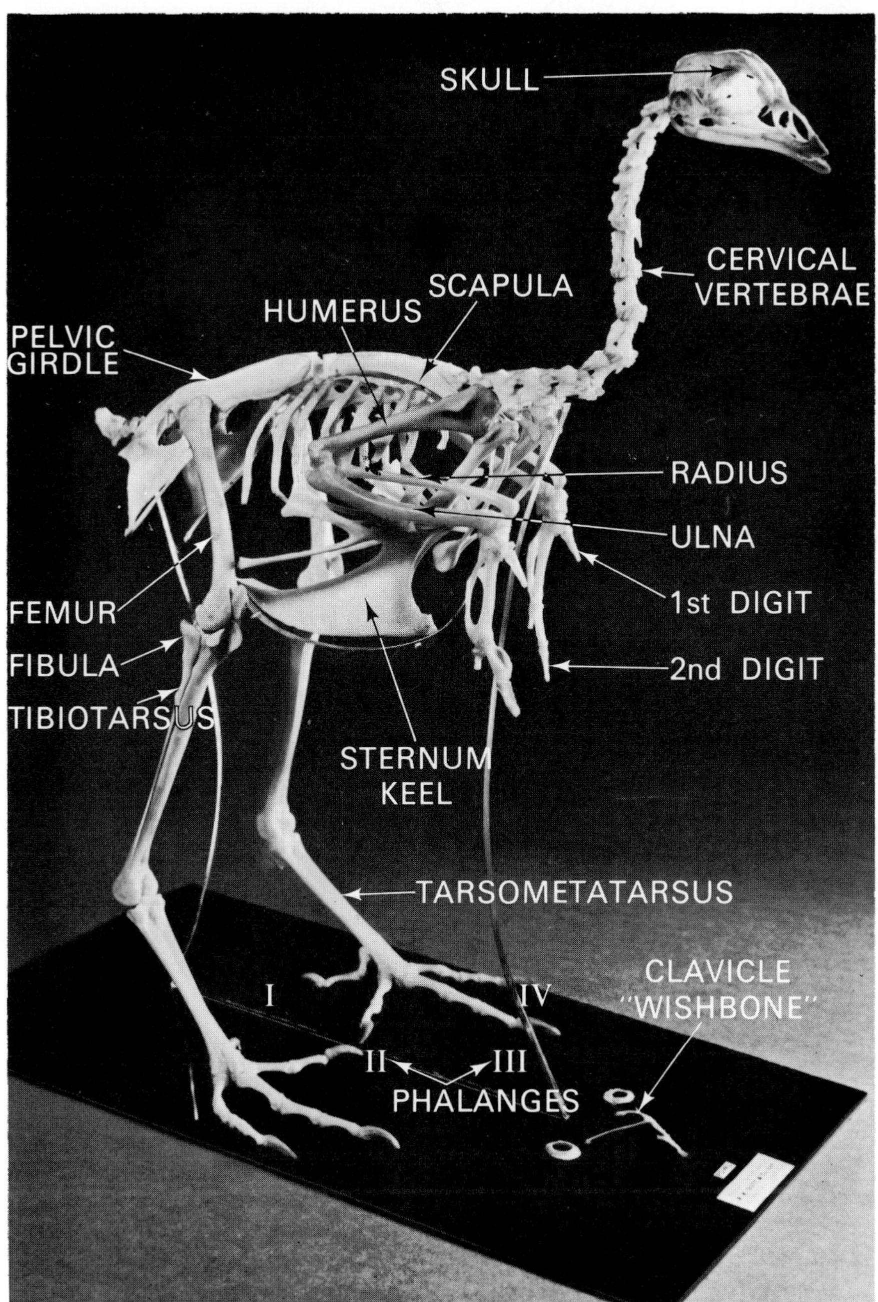

Plate 14 Skeleton of the domestic fowl (Griffin Biological Laboratories)

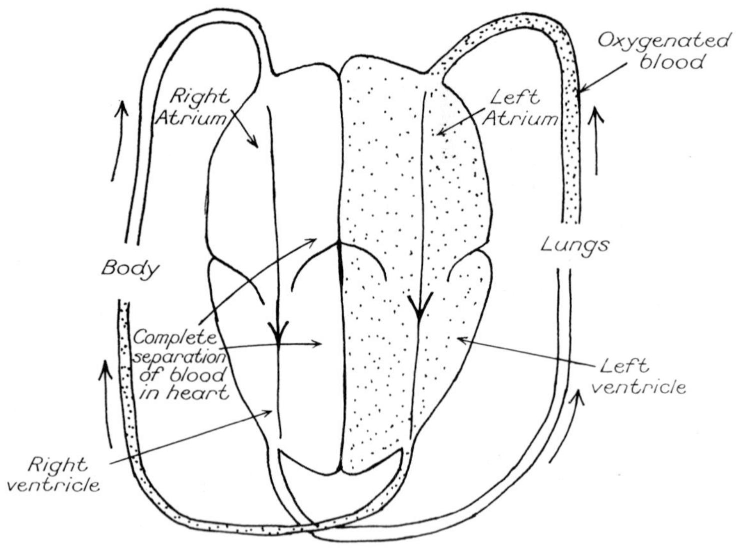

Fig. 4.14 Blood circulation in a bird's heart

CIRCULATORY SYSTEM

Dissolved food and oxygen are circulated by means of a *four* chambered heart composed of *two* auricles or atria, and *two* distinct ventricles, this is a structure closely resembling the heart in man. Oxygenated blood enters the left auricle and then is pumped by the left ventricle to the body cells through the *arterial* system. Deoxygenated blood returns to the right auricle to be pumped to the lungs by the right ventricle. There is a *complete separation* of deoxygenated and oxygenated bloods as shown in Fig. 4.14.

RESPIRATION

Rapid beating of the wings during flight demands a high energy supply in the breast muscles coupled with an efficient supply of oxygen.

Air is drawn into the nostrils in the beak, and passes down the *trachea* and *bronchi* to enter the lung with its good supply of blood capillary vessels. Air sweeps through the lung tubes to pass into the *air sacs* situated in the hollow bones and in spaces between the alimentary canal. The lungs in man have a stagnant or *residual air space* amounting to approximately 25% of the lung volume. In birds no such stagnant air space is found as air is drawn completely *through* the lungs into air sacs which hold air but *cannot* function as lungs in being able to exchange gases.

The process of inspiration and expiration is achieved by the *compression* of the body by the powerful chest muscles causing exhalation, whilst relaxation results in inhalation. When the bird is not in flight, *intercostal* muscles between the ribs are used in the breathing process.

Birds in contrast to fish, amphibia and reptiles are warmblooded or *homoiothermic* being able to keep a steady body temperature independent of the changing temperature of its surrounding environment.

REPRODUCTION

Fertilisation of the bird's egg is *internal.* The fertilised egg has the essential features to allow the development of the chick embryo to proceed until *hatching* after a prolonged period of *incubation* in which both parent birds may take part. *Parental care* of the young may continue for some time until the fledgling birds are able to feed and fend for themselves.

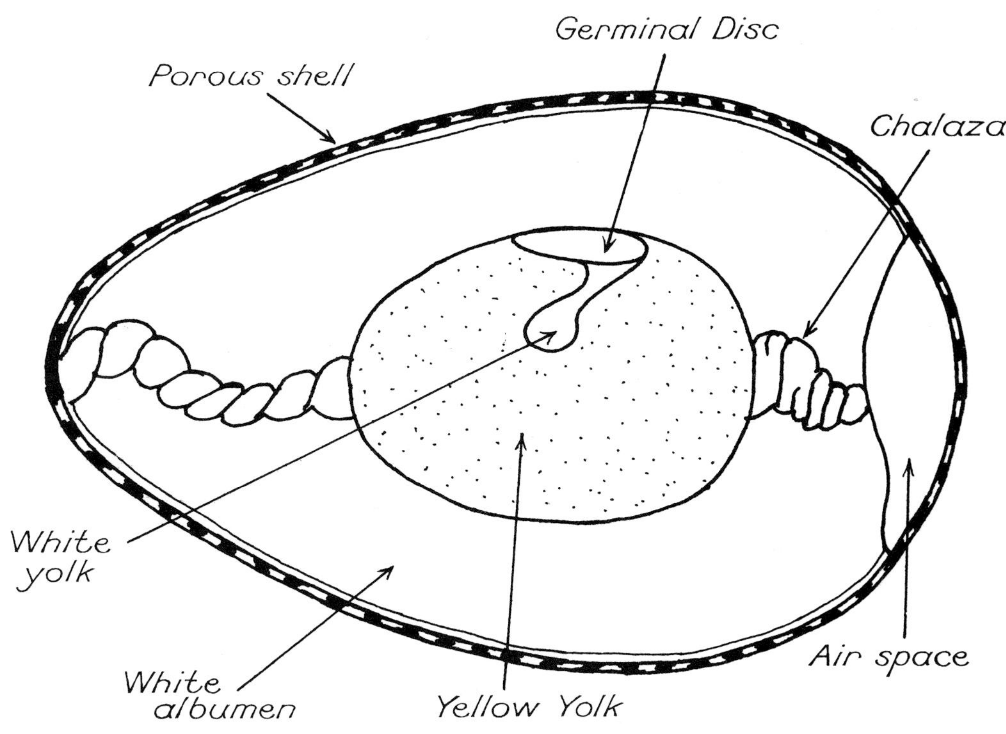

Fig. 4.16 Hen egg structure

The bird's egg has the following structure:

(a) **Shell.** This is a porous protective cover allowing air to enter the egg through the shell pores.

(b) **Germinal disc.** This marks the position of the future embryo and is located above the yolk which is a rich supply of food for the developing embryo.

(c) **Albumen** or egg white provides the watery environment essential for the development of land animals. It also protects and cushions the young chick from mechanical injury.

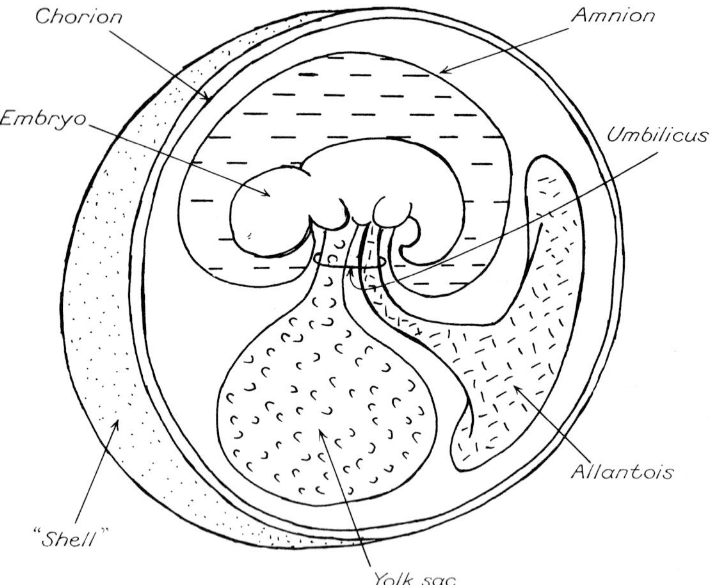

Fig. 4.17 Embryonic membranes of an egg laying animal

EMBRYONIC MEMBRANES

The cells of the germinal disc divide rapidly in an orderly manner by *mitosis* producing the *multicellular* embryo which proceeds to form the three important cell layers of the *ectoderm*, *endoderm* and *mesoderm*. Further division and development produces the *embryonic membranes* shown in Fig. 4.17.

(1) **Amnion** membrane completely surrounds the embryo enclosing it in a 'pond' of water or *amniotic fluid.*

(2) **Yolk sac** provide the food for growth of new cells and tissues from the *protein* present in the yolk sac directly connected to the alimentary canal of the chick.

(3) **Allantois** is a small sac serving a number of purposes, amongst which it collects excretory waste; serves to exchange oxygen through the well developed blood capillary network; and to absorb water from the egg white.

(4) **Chorion** is the outermost membrane completely surrounding the embryo and all other membranes, and is directly beneath the egg shell. In man the chorion is one of the links between the human embryo and the womb or uterus wall.

A short *umbilicus* is formed from the yolk and allantois stalks.

EXCRETION

The kidneys of the bird produce a highly concentrated form of urine in a semisolid form due to the removal of excess water retained by the body, this pasty excretory waste is eliminated with faeces by way of the cloaca.

5 Mammals and Man

Mammals are the group of vertebrates which have evolved along a different branch to that of birds; both groups are believed to have a common ancestor amongst the reptiles.

BODY STRUCTURE

The body of a mammal consists of a *head*, a distinct *neck*, and a *trunk* which is divided into a *thorax* and *abdomen* by an internal *diaphragm*. A *tail* may also be present.

The *skin* of the mammal forms partly from the ectoderm layer and always contains a few *hairs* growing out of skin pockets called *follicles*, each having a *sebaceous* gland producing an oily wax secretion to maintain the supplement of the skin and help to make it waterproof and prevent water loss by evaporation. *Sweat glands* secretion to maintain the suppleness of the skin and help to make *mammary* glands have the ability to secrete the fluid *milk*.

Keratin protein is the main component of skin scales, hair, finger nails, hooves and horns of mammals.

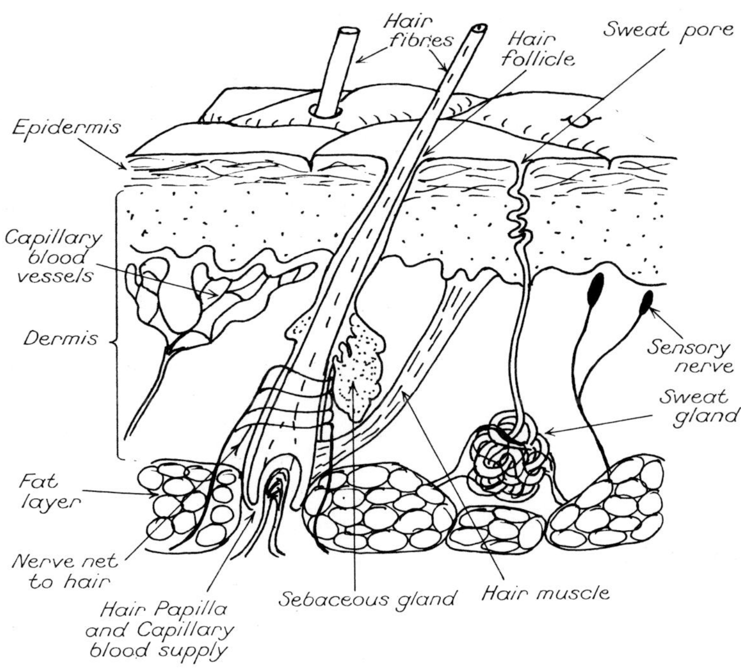

Fig. 5.1 Mammal skin structure seen in sectional view

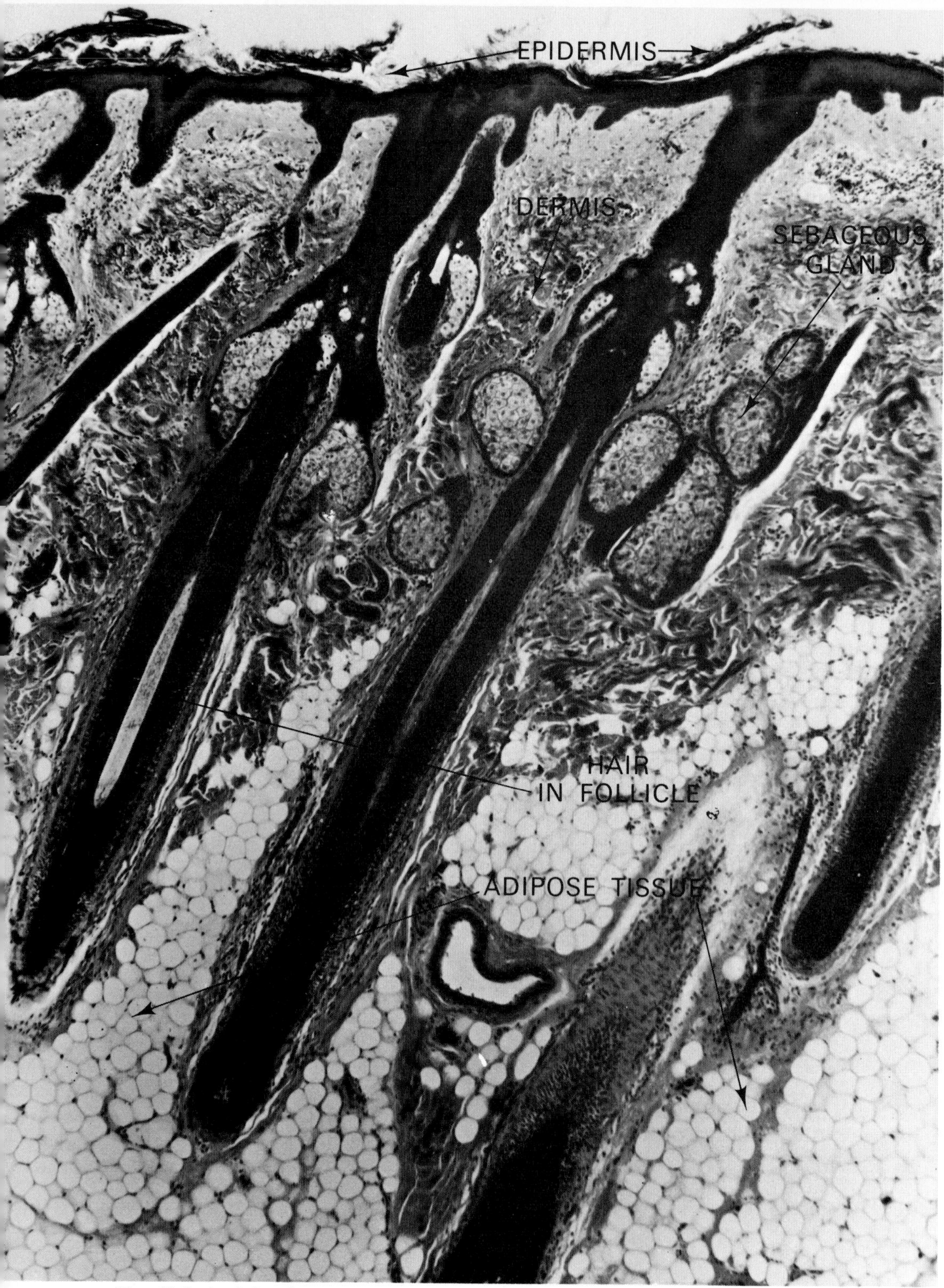

Plate 15 Sectional view of mamma skin with hair (Griffin Biological Laboratories)

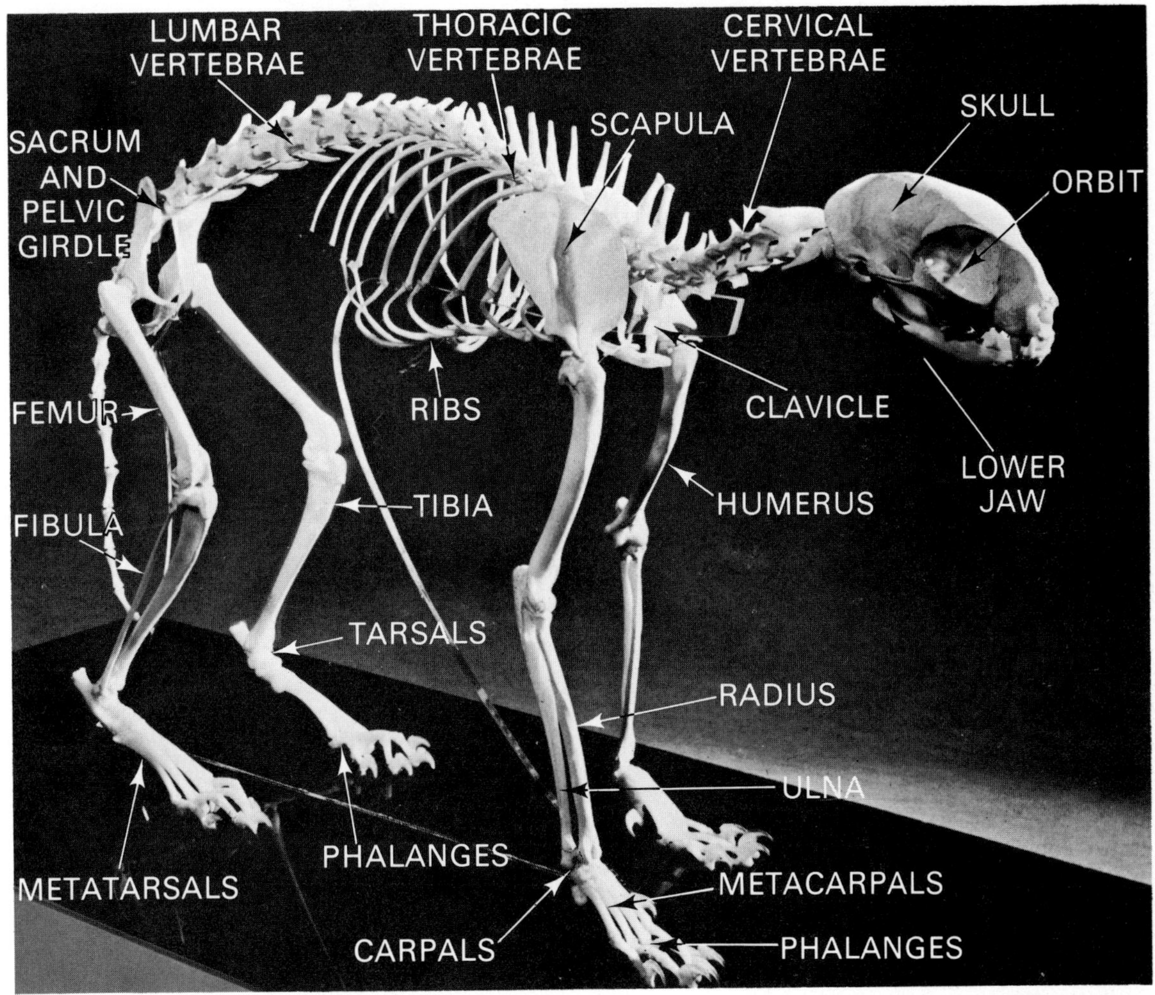

Plate 16 Complete skeleton of a cat (Griffin Biological Laboratories)

SKELETON

The same fundamental skeleton pattern is found amongst mammals, the individual bones being given the same names as in the skeleton of man. Modification of the different mammal skeletons is a result of adaptation to different modes of life, the main modification being in the appendicular skeleton and the pentadactyl limb (see Fig. 4.7, p. 50).

The *pectoral girdle* is loosely attached to the vertebral column, allowing the animal to land on its forepaws without causing serious injury other than *dislocation* or fracture of the clavicle collar bone.

The *pelvic girdle* is *strongly* bound to the vertebral column in a firm foundation of the *sacrum*. This strong connection is needed to withstand the driving force produced in propelling the body by the hind limbs.

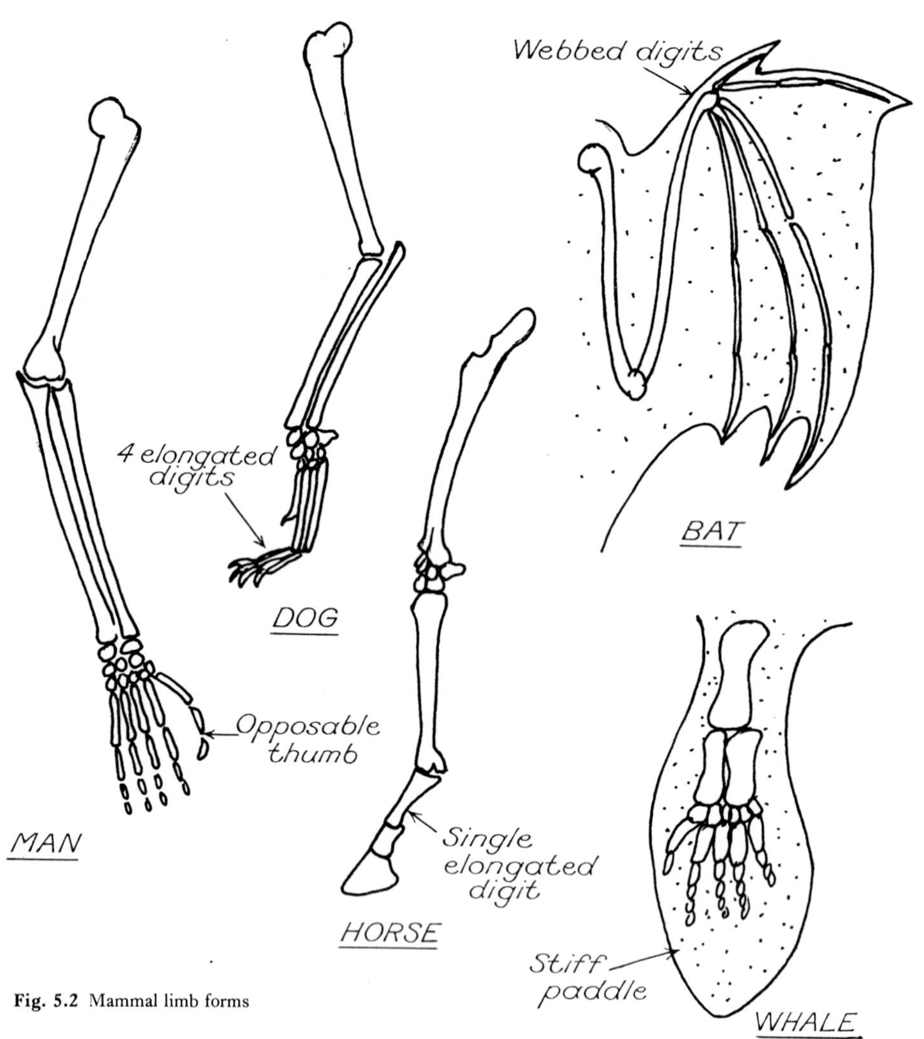

Fig. 5.2 Mammal limb forms

Modification of the forelimb of different mammals is shown in Fig. 5.2; see also Plate 13, p. 51.

NUTRITION

The skulls of mammals have a lower jaw which is *freely* movable and an upper jaw firmly *fixed* to the cranium. Each of the jaws carry two sets of teeth in succession called the *first* set or 'milk' teeth, and the *second* set or permanent teeth. Teeth of mammals are essential parts of the digestive system, being a masticatory apparatus serving to cut, shear or tear, and grind the food.

Incisors, are sharp, chisel-like teeth well developed in herb eating herbivorous animals, whilst the flesh eating *carnivorous* animals have well developed pointed *canine* teeth. The flat, crowned, grinding teeth are called *premolars* and *molars*. The basic anatomical structure of a tooth is shown in Fig. 5.3.

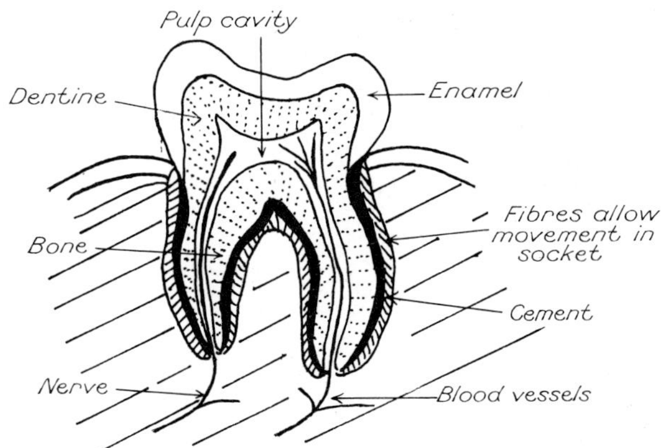

Fig. 5.3 Molar tooth structure

The food diet of different mammals is summarised:

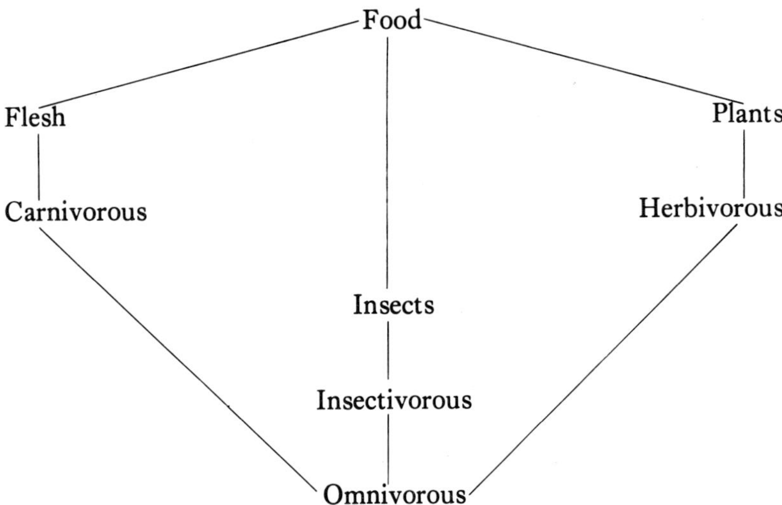

The secretion of the *salivary glands* of the mouth is mixed with the chewed food which passes along the *oesophagus* and the intestine by movement called *peristalsis*. Some digestion of food takes place in the *stomach* by action of gastric juice *enzymes*. Further enzymes are provided by the secretions of the liver and pancreas which enter the *duodenum*.

The intestine is in two parts: the *small intestine* is of considerable length and produces the *intestinal juice* as a secretion of the endoderm cell lining; it is also able to absorb the digested food. The *large intestine* is mainly for the purpose of extracting water and certain mineral salts from the faeces; it terminates in the *anus*, for the discharge of faecal waste. The urinary and genital openings are separate from the anus with its high microbe content in order to prevent infection of these normally microbe-free systems.

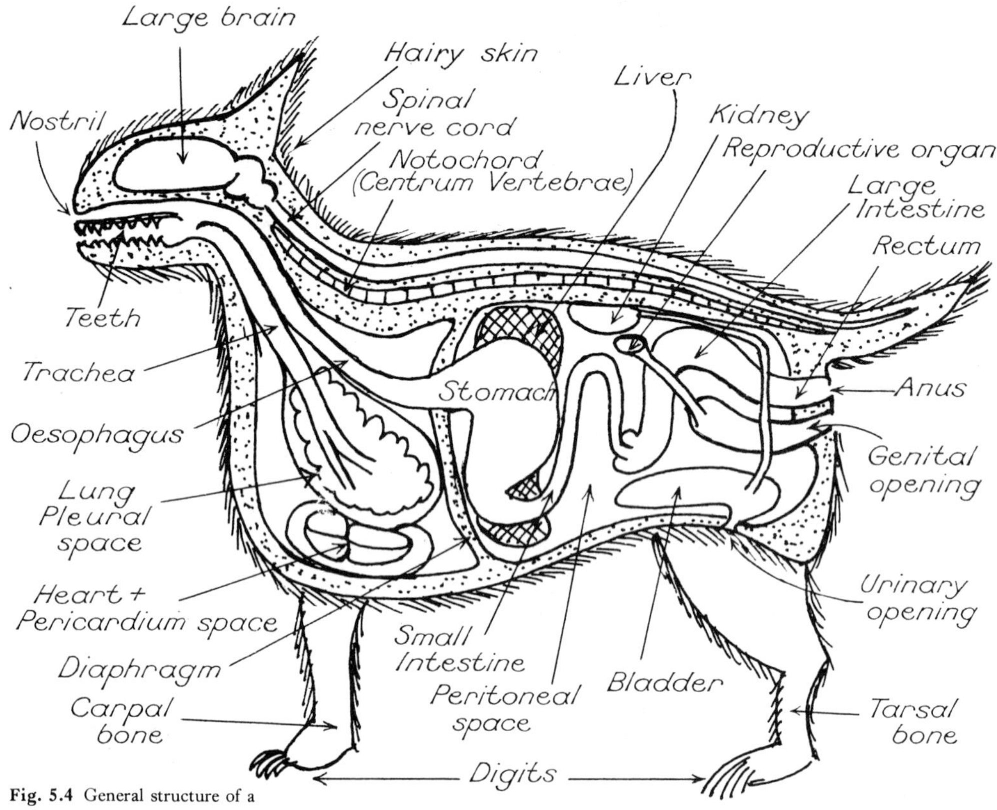

Fig. 5.4 General structure of a mammal

CIRCULATORY SYSTEMS

The mammalian heart is a *four* chambered structure of *two auricles* or atria and *two ventricles*, the latter are separated completely by an *interventricular septum*. The blood circulation circuit is shown in Fig. 5.5 (p. 70) where *no mixing* of deoxygenated and oxygenated bloods occurs, in contrast to the circulation seen in fish, amphibia and reptiles (see pp. 49, 56, and 60).

Mammalian blood consists of blood cells; the red cells of man show the complete absence of a definite *nucleus*, which is seen in the red blood cells of other vertebrates.

Blue babies. The amphibian heart with its *single* ventricle allows the deoxygenated and oxygenated bloods to mix at some time during the circulation. Certain human babies are born with a hole in the interventricular septum, as a consequence the hearts of the 'blue babies' will function in a similar manner to the amphibian heart with incomplete separation of oxygenated and deoxygenated bloods. This causes breathlessness and inability to undertake strenuous exercise together with the blue appearance of the skin. Such defects of the heart are repaired by surgery.

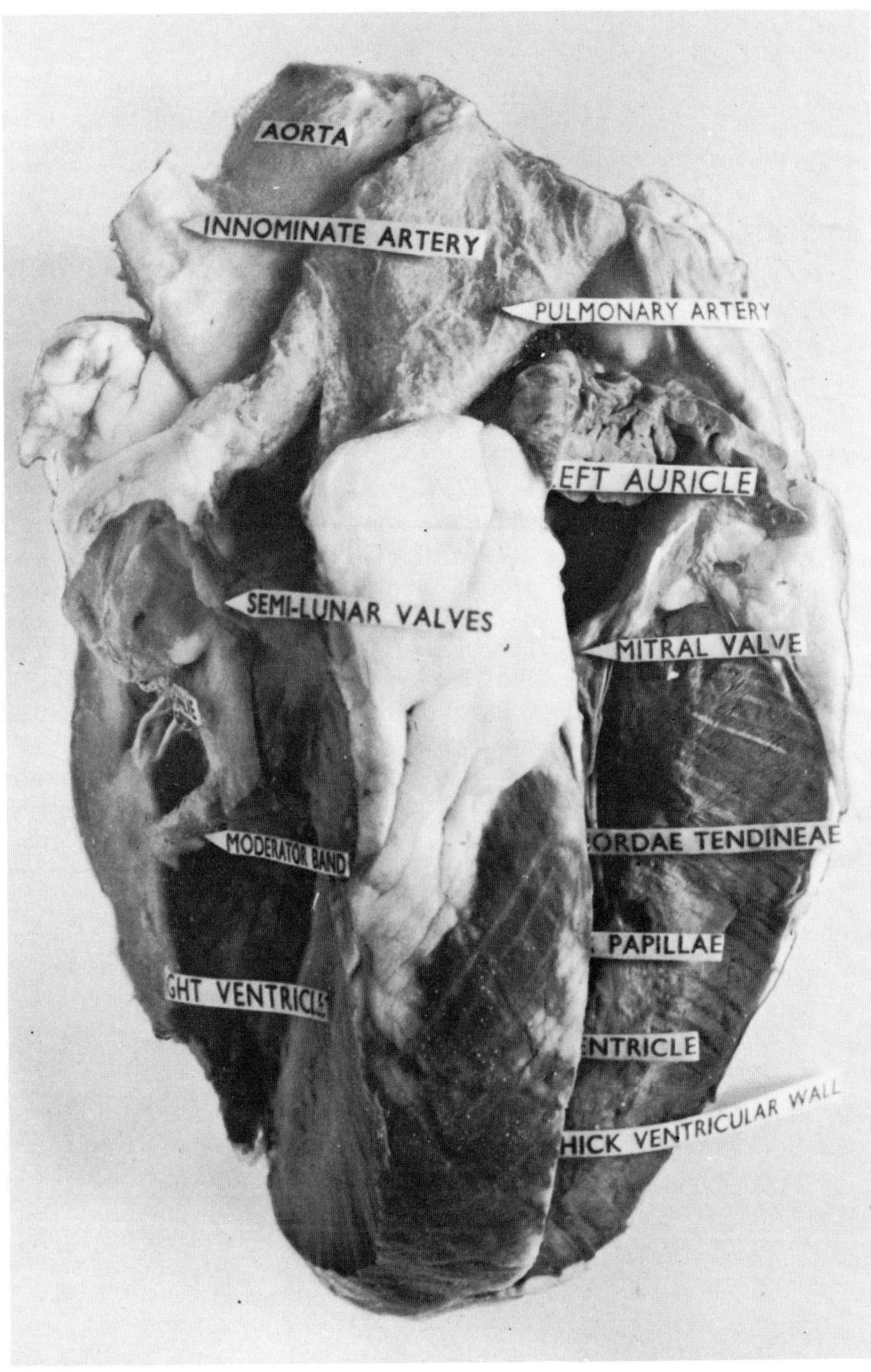

Plate 17 Dissection of a mammalian heart to show its structure (Griffin Biological Laboratories)

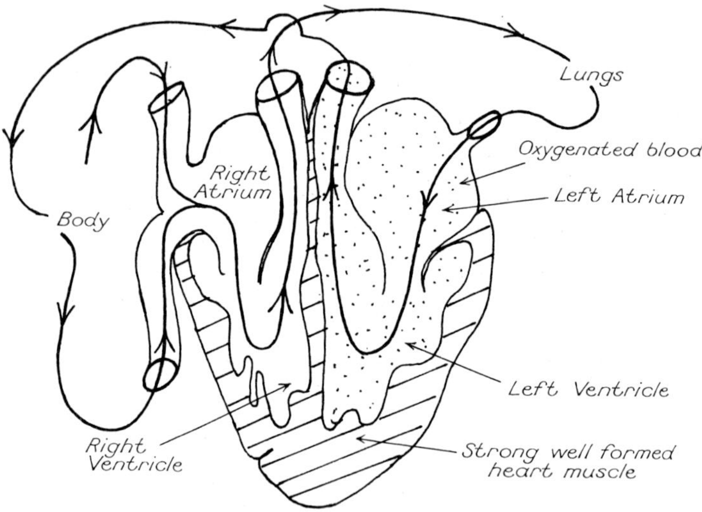

Fig. 5.5 Blood circulation in a mammal heart

RESPIRATION

Mammals draw air into the body through a system of air tubes called *trachea*, *bronchi*, and *bronchioles* which end in air sacs or *alveoli*, the latter have very *thin* walls which are continually kept *moist* and well supplied with a *capillary network* of blood vessels, this allows rapid exchange of gases by *diffusion*.

The *diaphragm* has an important function in the process of external respiration, the lowering of the diaphragm and raising of the rib cage, results in the air pressure inside the lung becoming *less* than the air outside which results in a rush of air inwards to the lung. During *exhalation* the diaphragm muscle and the rib intercostal muscles *relax* resulting in the pushing up of the diaphragm and collapse of the rib cage, causing air to be expelled from the lungs.

The method of external respiration contrasts with that used by other vertebrates, fish, amphibia, birds and reptiles, who inhale air

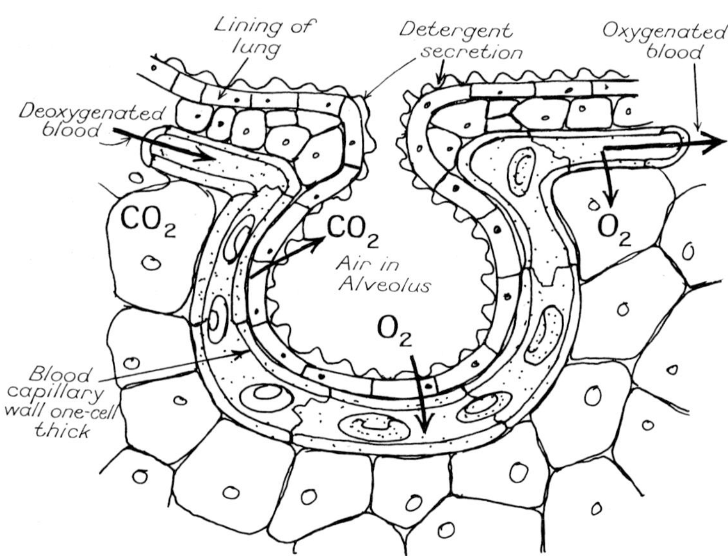

Fig. 5.6 Lung alveolus of a mammal

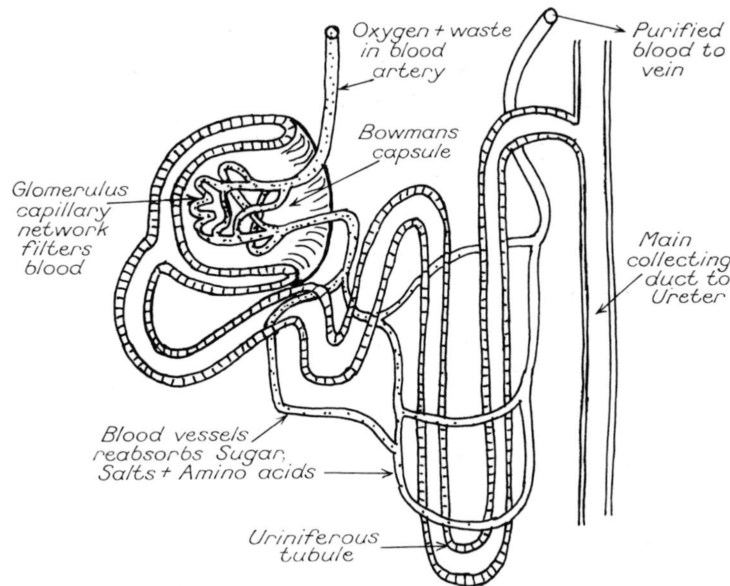

Fig. 5.7 Detail of kidney structure

mainly by ***relaxation*** of the body wall muscles, and exhale by ***contraction*** of body wall muscles.

Mammals are warm blooded or ***homoiothermic.***

EXCRETION

The ***kidneys*** are the main organs of excretion which eliminate water, salts, and urea (carbamide) by ***filtration*** of the blood by positive blood pressure through the glomerulus and Bowman capsule. ***Diffusion*** of salts and certain digested foods, glucose and amino acids takes place into the kidney tubules. ***Osmosis*** of water into the kidney tubules and blood vessels prevents excess water loss, whilst ***reabsorption*** of glucose and amino acids and certain salts also takes place into kidney tubules and blood vessels.

The urinary ***bladder*** is connected by ***ureters*** to the paired kidneys, which links in the female by the ***urethra*** to the exterior by a separate urinary aperture, whilst in the male the urethra is a combined outlet for urinary and reproductive products opening to the exterior in a ***urino-genital*** aperture.

NERVOUS SYSTEM

The vertebrate brain consists of the following main parts:

(a) *Cerebral hemispheres* or *cerebrum* concerned mainly with memory, intelligence; also located here is the area concerned with the sense of smell.
(b) *Optic lobes* concerned with sight.
(c) *Cerebellum* concerned with balance, hearing and the sense of touch.
(d) *Medulla oblongata* controlling the heart beat and the rate of breathing.

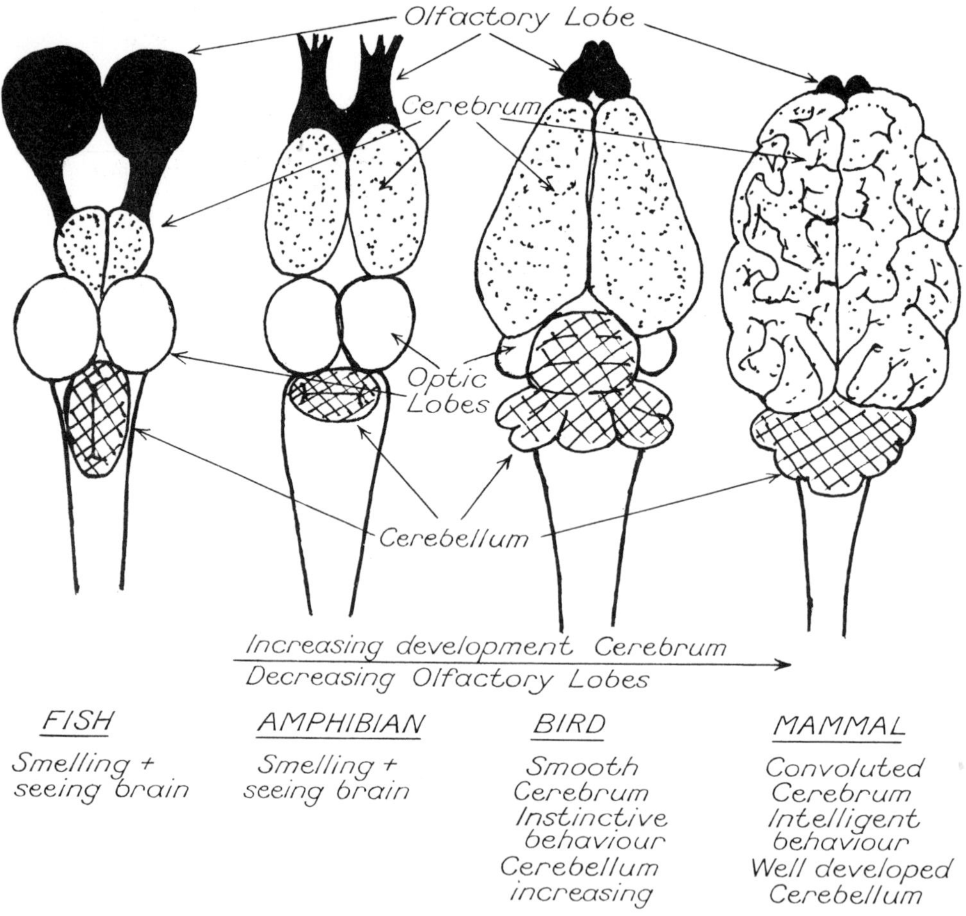

Fig. 5.8 Comparison of vertebrate brains

Figure 5.8 compares the brains of different vertebrate animals: a gradual enlargement of the cerebral hemispheres is seen from fish to mammals. In mammals the cerebral hemispheres increase their surface area and the number of brain nerve cells by means of in-foldings or convolutions seen to the greatest extent in the brain of man.

Cranial nerves connect the brain with different parts of the body; there are twelve pairs in man and mammals compared to only 10 pairs in other vertebrates. The additional cranial nerves in mammals are the:

11th Accessory to the neck and thorax.
12th Hypoglossal to the muscle of the tongue.

Spinal nerves from the spinal cord to the arms, legs and trunk, indicate the *segmental* nature of the vertebrate body.

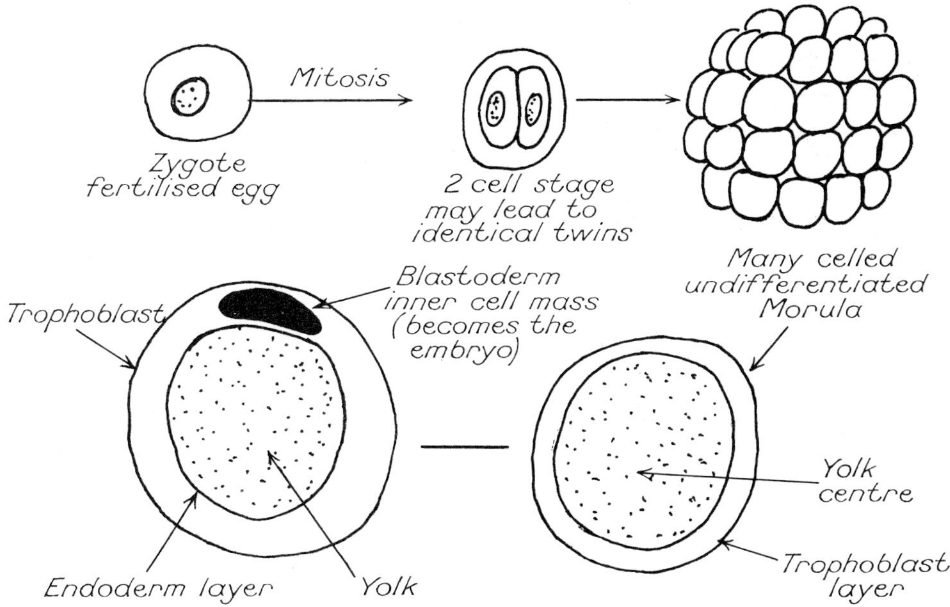

Fig. 5.9 Early development of a mammal zygote

REPRODUCTION

The male and female sexes are separate. The male system consists of a pair of *testes* producing *spermatozoa* which pass into the *epididymis* to enter the *urethra* within the *penis*. The female system consists of a pair of *ovaries* which shed the *ova* into *oviducts* connecting with the *uterus* and *vagina*.

Fertilisation in mammals is *internal*: the sperm are ejaculated from the penis which is inserted into the vagina during *copulation*. The *seminal fluid* provides the essential watery medium for the spermatozoa to swim in, and also provides food in the form of soluble glucose needed for energy for the vigorous swimming movement of the spermatozoa in the oviducts of the female. On reaching the ova, fertilisation by a single spermatozoon occurs to produce the *zygote*, which becomes surrounded by an envelope of albumen. Instead of becoming surrounded by a shell, the fertilised mammal egg is *implanted* in the lining of the uterus.

Development of the embryo mammal is *internal* within the uterus of the female; this provides a constant environment of warmth, moisture and food supply within the close protection of the uterus. The length of time which the embryo spends developing in the uterus is called the period of *gestation* and differs for different mammals.

[A kangaroo allows the embryo to complete its early development externally in the pouch, whilst the duck-billed platypus is the only mammal laying its fertilised ova in shells.]

As the rapid nuclear division, by mitosis, of the zygote nucleus proceeds (see Fig. 1.11, p. 12) the outer cells become the special *trophoblast* layer, which burrows into the uterus wall when the fertilised egg is implanted. Body cells develop from a special group of cells called the *inner cell mass* which become the primary *ectoderm, endoderm* and *mesoderm* cell layers.

Embryo membranes develop with continual growth of the embryo; these membranes are essential for internal development to take place.

(a) The *amniotic membrane* surrounds the embryo with an amniotic fluid providing the water of the 'pond'.

(b) A *yolk sac* develops as an extension of the alimentary canal of the embryo and serves only for a short time since the feeding of the mammal embryo is from the mother.

(c) An *allantoic bladder* similar to that seen in the chick embryo, assists in the feeding of the mammal embryo, but in man the allantois has no function and is an example of a *vestigial organ.*

(d) The *chorion* completely surrounds the embryo and the other membranes, that is the amniotic and allantoic bladders and yolk sac.

(e) The *placenta* provides the close union between the embryo and the uterus wall. In man the placenta is formed from the *chorionic membrane*, whilst in other mammals the *allantoic* membrane may be involved.

The *umbilical cord* consists of the stalks of the yolk sac and allantoic bladder, together with blood vessels from the placenta.

The main functions of the placenta are to:

(1) Transfer water, food, and oxygen to the embryo.

(2) Remove waste from the embryo.

(3) Protect the embryo from harmful microbes, but some such as the German measles virus can be passed from the mother, as well as harmful drugs and chemicals.

(4) A hormone called human chorionic gonadotrophin (HCG) is produced by the chorionic membrane and serves to prevent further ova developing.

Coelomic spaces develop in the middle of the mesoderm layer of cells in the embryo. These spaces persist and are seen in the adult mammal as the *peritoneal cavity* of the abdomen, the *pleural cavity* surrounding the lungs, and in the *pericardial cavity* surrounding the heart. All of these are lined with a *serous* membrane.

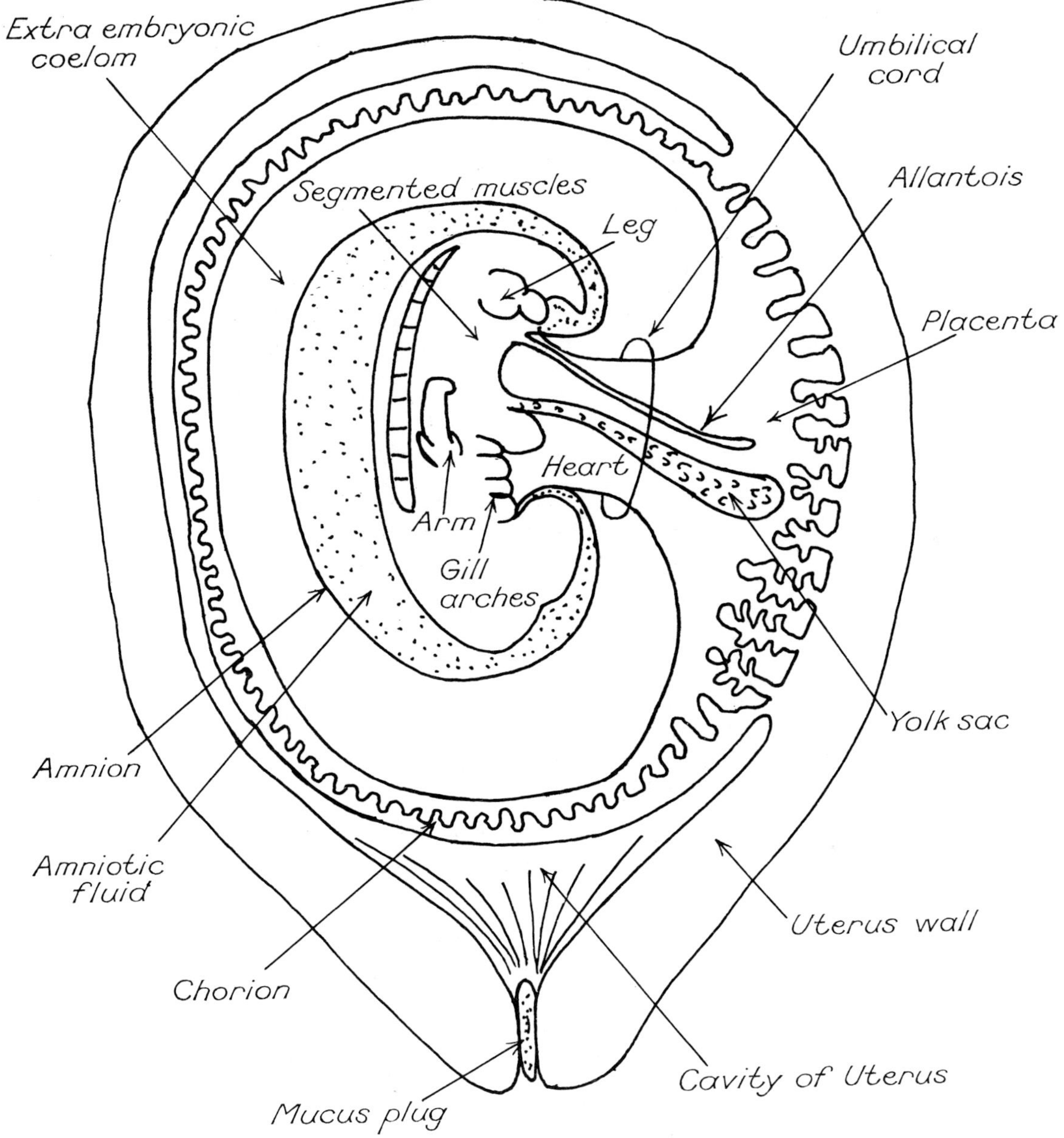

Fig. 5.10 Embryo membranes of a mammal

Parental care is given to the young mammals after their birth, in a period of *suckling* the young on milk fed from the mammary glands, and during *weaning* to more varied foods. Parent mammals continue the process of care and protection with that of teaching by example until the young are able to fend for themselves. Parental care reaches its highest and lengthiest state in man.

Plate 18 Neanderthal man, a prehistoric man who lived in Europe 200 000 years ago (Bayer Chemicals Ltd)

MAN AND THE MAMMALS

Man is related to the tree dwelling or *arboreal primates* which include the apes, the monkeys, these mammals lead a mode of life in the branches of trees away from prowling carnivorous tigers and wolves. Several features distinguish the *primates* from other mammals.

(1) They walk on the soles of their feet or the palms of the hand: this is called *plantigrade* walk.

(2) Hands and feet are adapted to *grasping* branches: the thumb and big toe are set differently to the other digits and are said to be *opposable*, in order to grip.

(3) The sense of sight involves a change in which the eyes are *forward looking*: this produces *binocular* vision which allows primates to judge distances, important when leaping from branch to branch.

(4) The brain shows a greater development in the *cerebral hemispheres*.

FEATURES OF MAN

Modern man called *Homo sapiens* is placed apart from mammals and other primates in the following:

(a) *Speech* as a means of communication.

(b) *Bipedal* mode of walking on two feet, the body upright, and the big toe is *not* opposable to the other toes.

(c) The hand of man can grip in two ways: the *power* grip for holding hammers and clubs, a grip shared by other primates; and the *precision grip* seen when holding small instruments such as a pencil when writing—this grip is not developed in any other primate.

(d) The *cranium* of man has increased in *volume* to accommodate the cerebral hemispheres which have increased in size.

(e) *Hair* is absent from much of the body surface of man, compared to the primates.

MAN'S ANCESTORS

Evidence from fossils are used to trace the ancestry of modern man, through a study of the things primitive man could make or in *artifacts*, and by a study of the remains of his *home*, also give further indications.

Pre-human or near man was a slow walking or ambling person, short and half stooping, who lived in the open pasture feeding on plants and small animals, omnivorously. His cranium size was about 500 cm^3. He could be considered as a man-like ape.

Early or true man, or Homo erectus, also called Java man, was a person with a greater ability to walk upright. The cranium volume was between 700 and 1000 cm^3. This early human being could use a power grip for handling clubs and stone axes and is believed to have been able to make and use fires.

Modern man, or Homo sapiens first appeared as neanderthal man who was an erect walking, strong legged person, with a cranium volume between 1000 and 1600 cm^3.

Neanderthal man could use his hands for power grips, and could possibly use them for a precision grip. He made many tools in stone and bone, and could also make fire and use it for cooking. He lived mainly in caves and was known to bury his dead relatives.

Cromagnon man had a striding walk and had power and precision grips; he worked with stone, bone, wood and different metals. He lived mainly in *communities* in lake and hill forts, being able to *converse* freely with those with whom he lived.

From modern man the existing *races* of mankind have developed namely the *caucasian, mongolian, negroid* and *australasian.* They differ mainly in their skin colour, hair texture, facial features, and show different blood groupings.

6 *Variety in Plant Life*

The plant kingdom includes those living things which have the following outstanding features, compared to the members of the animal kingdom:

(1) A *fixed*, sessile, or rooted way of life in either soil or water.
(2) The *body form* is spreading, or *branched*, and seldom compact.
(3) Rigid *cellulose* is the strengthening and supporting material of cells.
(4) *Light* is the source of *energy* for the majority of plants containing the coloured pigments of chlorophyll.
(5) *Simple raw materials* such as water, carbon dioxide and mineral salts are used in food manufacture.

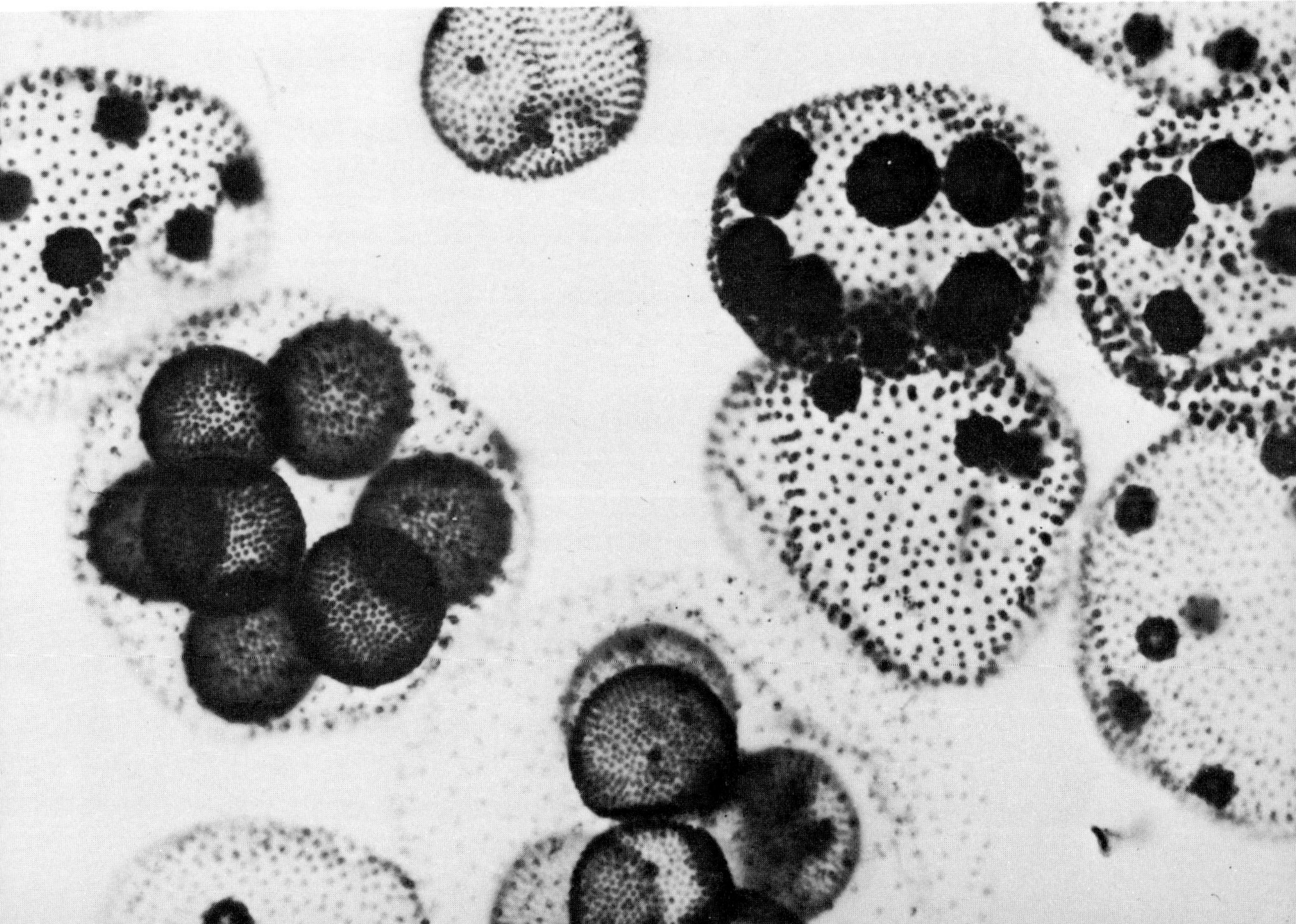

Plate 19 Colonies of the green algae, Volvox, found in pond water (Griffin Biological Laboratories)

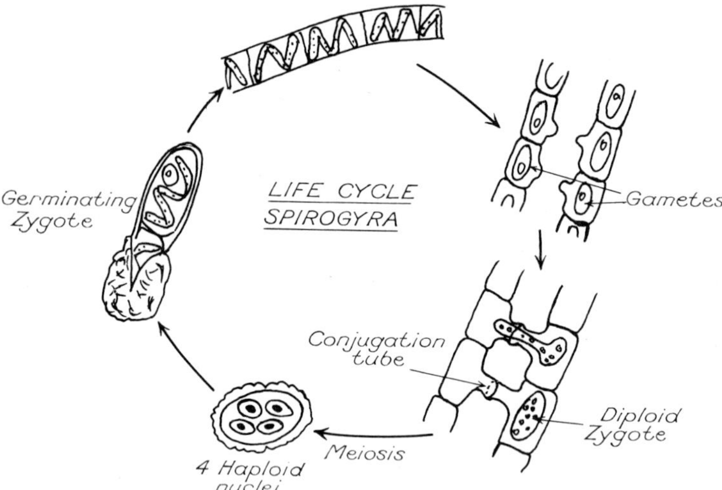

Fig. 6.1 Spirogyra life cycle

The following outline classification is given for the plant kingdom:

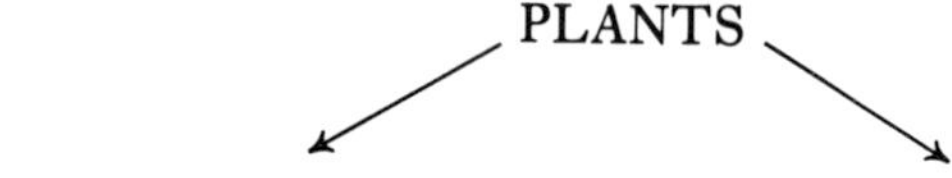

Nonflowering or SEEDLESS plants

(1) BACTERIA and single-celled bluegreen plants

(2) ALGAE single and multicellular green plants, chlamydomonas and seaweeds

(3) FUNGI, without chlorophyll, moulds and mushrooms

(4) LICHENS, combined algae and fungi plants

(5) LIVERWORTS, flattened green plants of damp places

(6) MOSSES, simple leafy plants

(7) FERNS, fringy leaved plants with roots

SEED producing plants

(8) NAKED SEED, or gymnosperm conifers, e.g. pine

(9) COVERED SEED or angiosperm plants, seed enclosed in a fruit

(a) *Monocotyledon*, or the narrow-leaved plants, e.g. cereals and grasses, tulip and daffodil

(b) *Dicotyledon*, or broad-leaved plant, e.g. the beech and rose

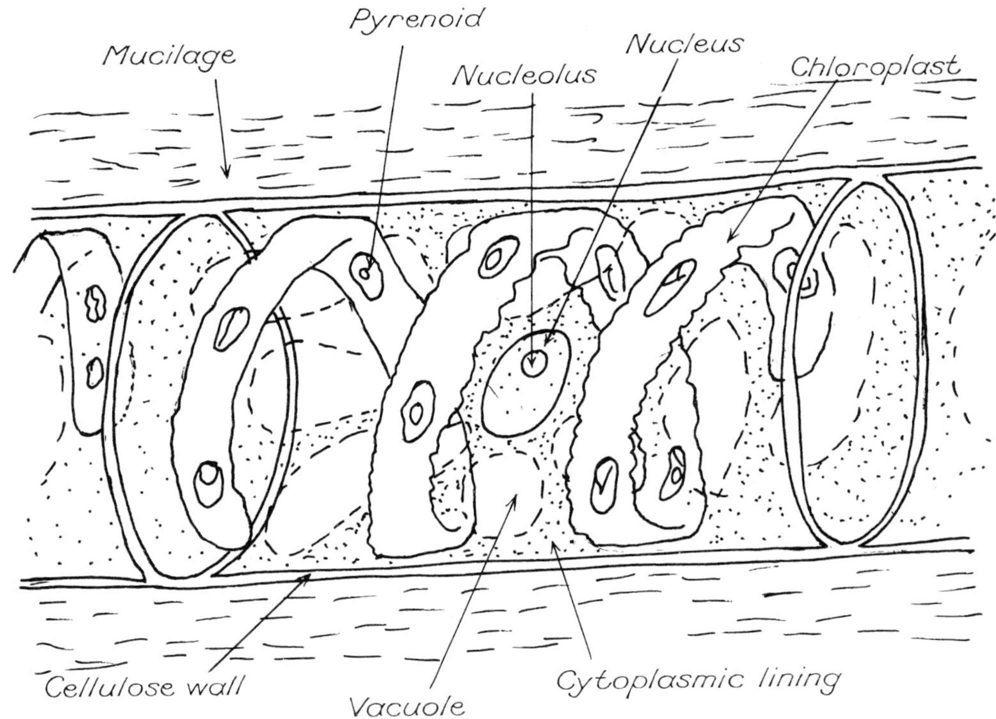

Fig. 6.2 Spirogyra cell structure

ALGAE: SIMPLE GREEN PLANTS

This group of simple, green coloured, plants includes the *unicellular* plants such as *Chlamydomonas* and the massive *multicellular* seaweeds.

BODY STRUCTURE

Algae are plants which do *not* possess roots, stems or leaves seen in most flowering plants. The plant body of algae being called a *thallus*, which can be a *single cell* in many algae, or *multicellular* in seaweeds composed of a flattened *frond*, thick stem and holdfast fixing the plant to the rocks.

REPRODUCTION

Apart from simple division of the cells by mitosis, the algae reproduce *sexually* by means of *gametes* which may be liberated into the surrounding water by seaweeds where external fertilisation occurs to produce a *zygote*; or adjacent filaments conjugate as in *spirogyra*. The life cycle of *spirogyra* is summarised in Fig. 6.1.

Freshwater algae

Typical freshwater algae are *Chlamydomonas*, *Volvox*, the *Spirogyra*, the latter consisting of a filament of individual cells joined end to end. The structure of a *Spirogyra* cell is shown in Fig. 6.2.

Masses of floating *Spirogyra* are supported by millions of bubbles of *oxygen* gas; this is an important byproduct of *photosynthesis* and serves a useful purpose in aerating water of the pond and returning oxygen to the air.

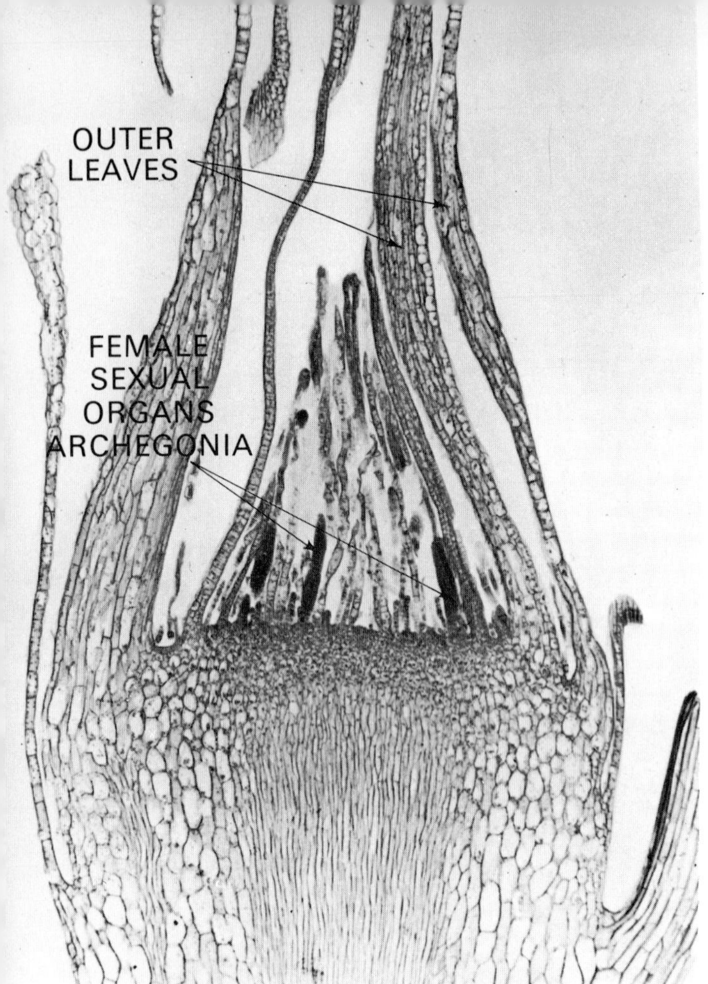

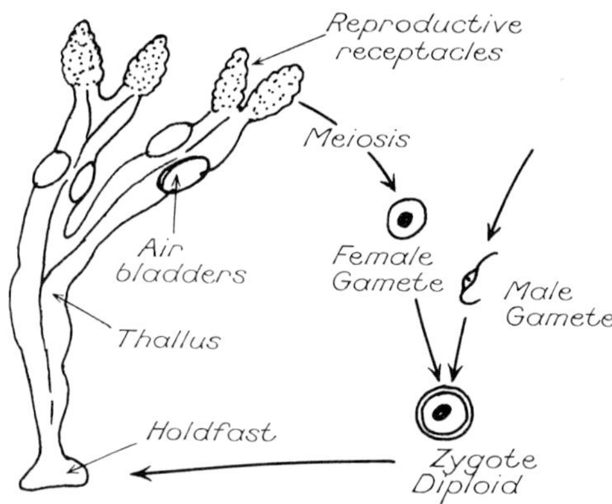

Fig. 6.3 Seaweed life cycle

Plate 20 The female sexual organs of a moss plant seen through the microscope (Griffin Biological Laboratories)

Marine algae

Seaweeds are of three colours: red, brown and green. The *red seaweeds* produce *agar* important for making the culture jelly used in the laboratory for growing microbes. *Brown seaweeds* are amongst the commonest algae on the seashore are used to produce *iodine* and the *alginate* salts used in the production of absorbable surgical dressings. *Green seaweeds* have a high mucilage content and *Carrageenan moss*, a seaweed, is used in medicinal and food preparations.

Plankton

Unicellular algae called *phytoplankton* form the food of herbivorous tiny shellfish or *zooplankton*. This is the commencement of an aquatic *food chain*.

MOSSES AND LIVERWORTS—BRYOPHYTA

Mosses and liverworts are the green coloured plants found living on land, in essentially wet and damp places.

STRUCTURE

Liverworts have a *thallus* or plant body which is flattened like a seaweed frond and fixed into the wet soil by tiny *rhizoids*.

Mosses have an *upright* thallus consisting of a central stalk surrounded by many *leaf-like* structures, a purpose of which is to manufacture food and hold water.

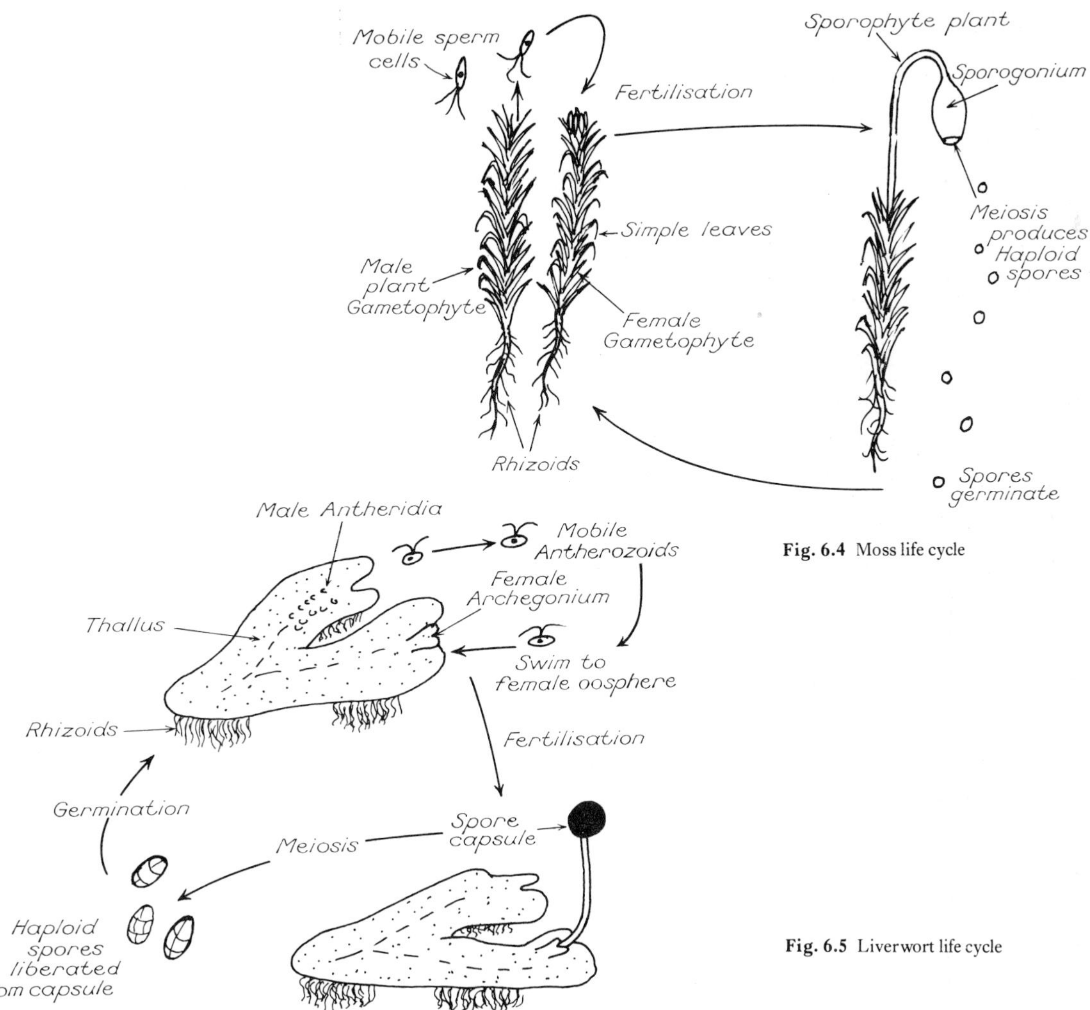

Fig. 6.4 Moss life cycle

Fig. 6.5 Liverwort life cycle

REPRODUCTION

Mosses and liverworts reproduce *sexually* by means of male reproductive organs called *antheridia*, producing *antherozoids* and female reproductive organs called *archegonia*, producing *oospheres*. The plant body which possesses these reproductive organs is called the gamete plant or *gametophyte*. It is necessary for the male antherozoid to swim in the water, present in abundance, towards the female oosphere to form the zygote.

The zygote of liverworts and mosses grows into a second plant called the *spore producing plant* or *sporophyte*. This consists of a *sporogonium* or spore capsule supported on a stalk above the thallus; it releases many tiny *spores*, each of which is carried by wind to land on suitable wet soil and grow into a new thallus or plant body of a liverwort or moss.

This plant *life cycle*, which included two distinct plant forms, spore producing *sporophyte* and gamete forming *gametophyte* is called an alternation of *generations*.

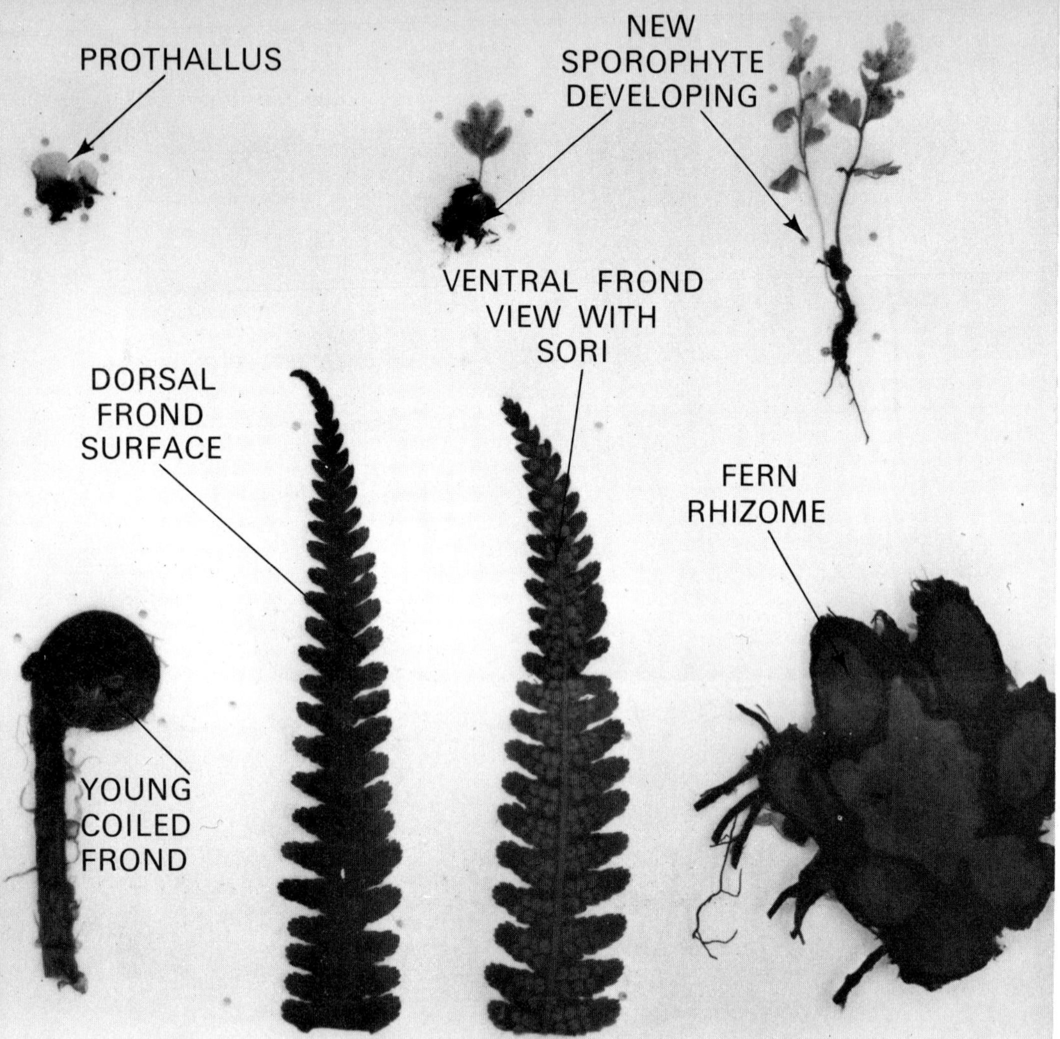

Plate 21 Stages in the life cycle of a fern (Griffin Biological Laboratories)

FERNS

Cool, damp places are favoured by most ferns, yet many can grow in conditions favourable to most flowering plants.

STRUCTURE

The fern thallus consists of a large broad, fringed shoot called a *frond*, arranged in such a way to get the greatest benefit of the sunlight.

Vascular tissue consisting of *tubular vessels* are seen inside the frond carrying water and salts to cells of the fern thallus; this development of special circulating *vascular tissue* is a consequence of the increasing size of the plant body.

REPRODUCTION

Behind the fern frond are groups of *spore* producing *sporangia* grouped together as a *sorus*. The fern frond is therefore the *sporophyte* generation of the plant.

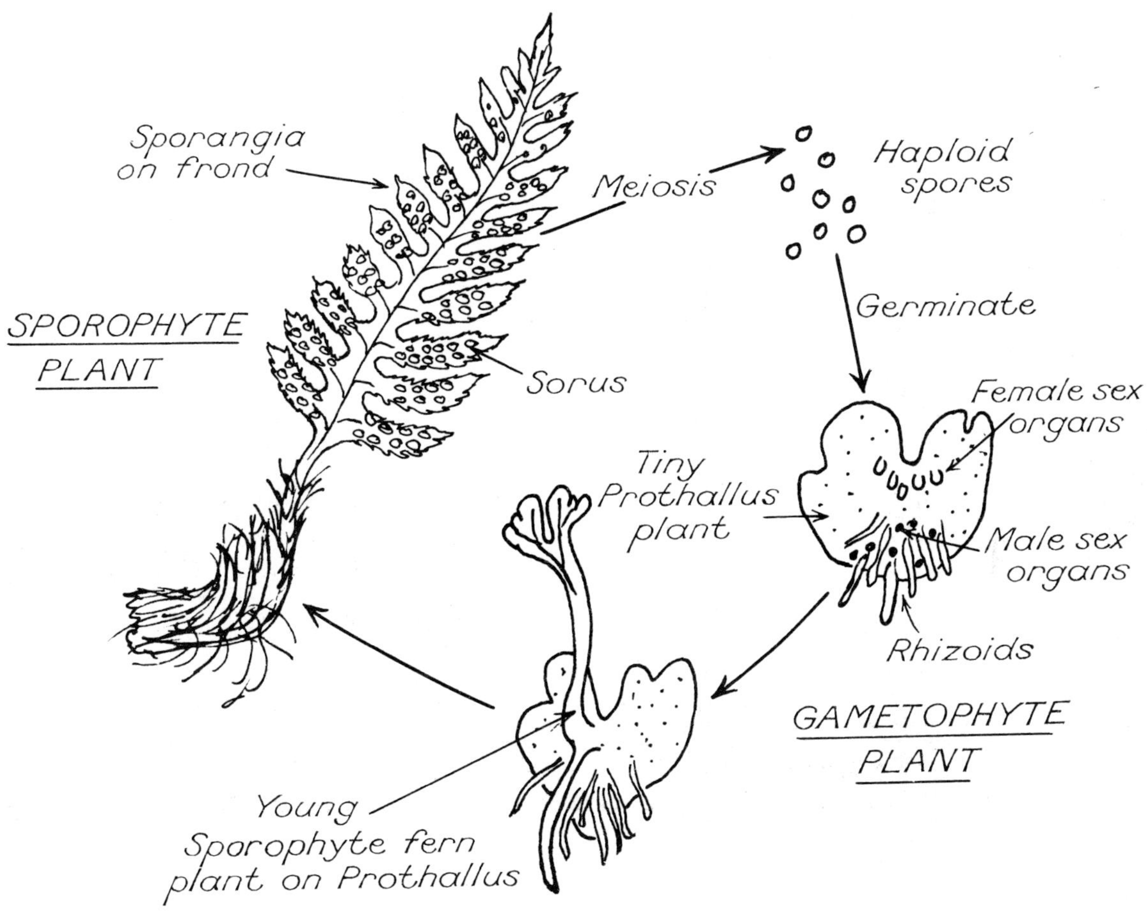

Fig. 6.6 Fern life cycle

Each tiny spore can germinate on wet soil to produce a completely *separate* miniature plant called a *fern prothallus*. The prothallus produces the *gametes* in the special sex organs. The male gametes then swim in the water covering the prothallus towards the female gametes and fertilise them to produce the zygotes.

The zygote germinates into the large frond of the fern plant thus producing the spore forming generation.

MEDICINAL USES

An extract of *male fern* is used for the displacement of the tapeworm parasite from the intestine of man.

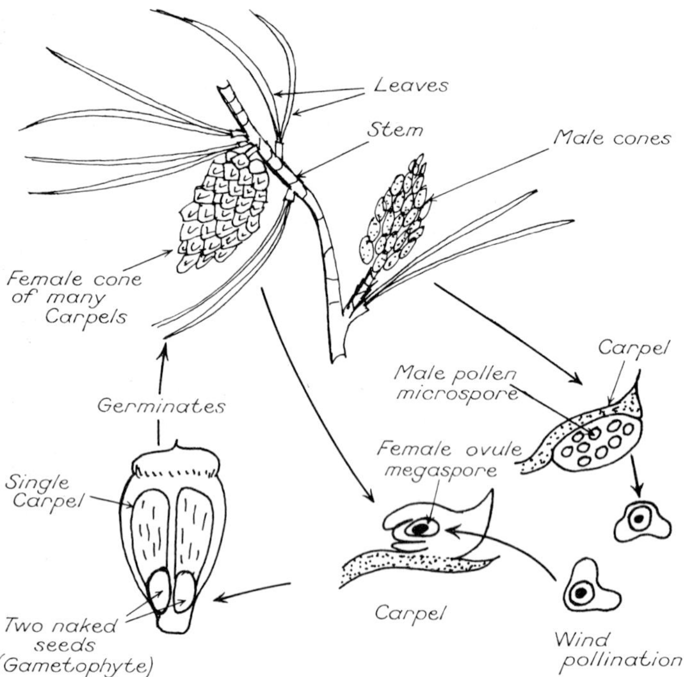

Fig. 6.7 Conifer life cycle

CONE PRODUCING PLANTS WITH NAKED SEEDS—CONIFERS

The *evergreen* conifer trees, *pine* and *spruce*, retain their leaves throughout the year, whilst the conifer *larch* is *deciduous*, losing its leaves in the autumn.

The most important structural feature of the conifers is the division of the plant body into *roots*, *stems* and the needlelike *leaves*; the possession of these structures is an advance on the simple plant bodies of ferns, mosses, liverworts and algae. In addition the great number of cells forming the conifer tree are linked with the greatly developed *vascular* system consisting of special tubular vessels. Amongst the vascular tissues are special *resin secreting* cells, producing the pleasant smelling oil of pine and also turpentine.

REPRODUCTION

Two types of *cones* are found on a typical conifer tree, a *male* and *female* cone. These cones are *spore* producing organs: the male produces *microspores* or *pollen grains*; the female produces *megaspores* located on the cone leaves.

Wind is responsible for transport of the microspore pollen, in contrast to the need for water to transport the male gametes in algae and ferns. The pollen enters the opening cones of the female fertilising the megaspore to produce the *zygote*. This develops into a miniature plant of *embryo* made of tiny leaves or *cotyledons*, a future root or *radicle*, and future shoot or *plumule*.

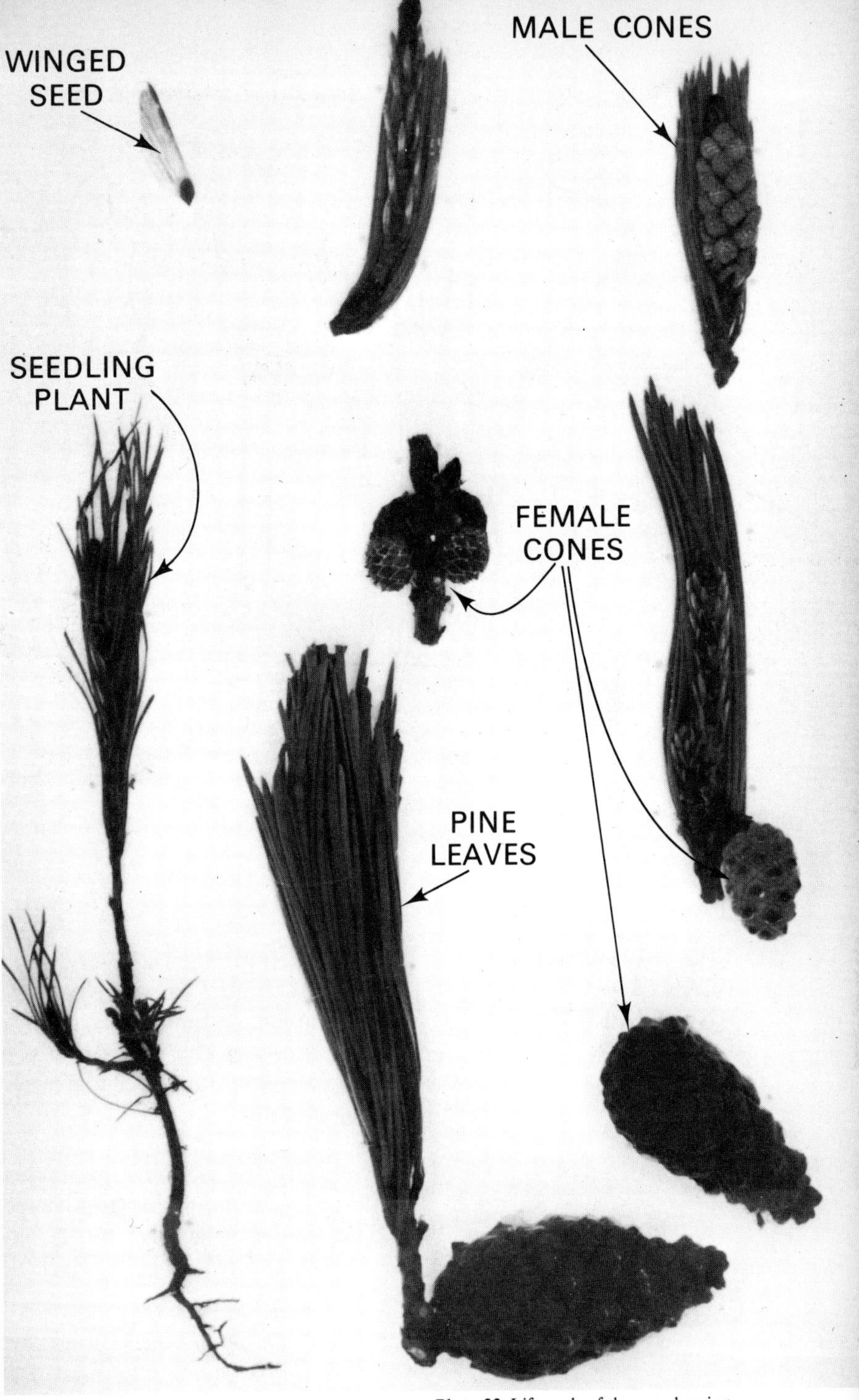

Plate 22 Life cycle of the cone bearing plant, the pine (Griffin Biological Laboratories)

Seeds of conifers are surrounded by a *testa* or seed coat, shaped into a wing which assists the dispersal of the seed by wind. Conifer seeds are '*naked*' in that they are found on the cone leaves and are not surrounded by a fruit wall seen in flowering plants.

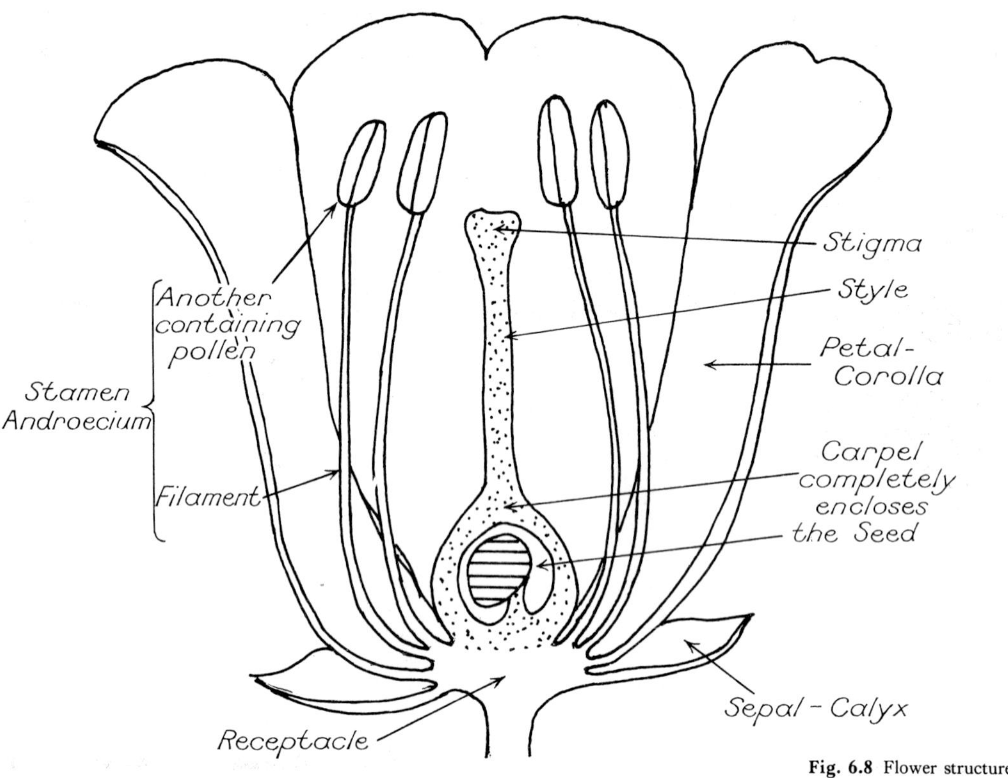

Fig. 6.8 Flower structure of an angiosperm

THE FLOWERING PLANTS—ANGIOSPERMS

Plants which produce *flowers* in addition to having a plant body of *roots*, *stems*, and *leaves*, are considered as the higher plants.

A complex system of *vascular vessels* carries water and salts from the roots to the leaves, whilst foods made in the leaves, and all chloroplast containing cells, are transported to the roots and other storage organs.

REPRODUCTION

Flowers are the *sexual reproductive* parts of the plant, whilst all other parts, roots, stems and leaves are the *vegetative* parts.

The flower is similar to the cones of conifers, but differs in possessing colourful *petals*. Each flower is composed of four main parts:

(a) *Calyx*, or *sepals*, protecting the flower in bud.
(b) *Corolla*, or *petals* which are coloured and sometimes have nectaries, and may be scented.
(c) *Androecium*, or *stamens* which produce the male pollen in pollen sacs supported on filaments.
(d) *Gynaecium*, or *carpels* the female part consisting of an ovary containing the *ovules*. Connecting with the ovary are the *stigma* and *style*. All the parts are attached to a flower stalk or *receptacle*. See Fig. 6.8.

Many flowers have both the male and female flower parts in one and the same flower and are *bisexual* or *hermaphrodite* flowers. A few flowers are *unisexual* having either a male androecium, or female gynaecium.

POLLINATION

This is the act of transferring pollen in flowers from the stamens to the stigma, by means of either *wind* or *insects*. It is essentially a process suited to dry land. Many flowers are *cross* pollinated with pollen from other flowers; others are *self* pollinated, the stamens providing pollen for the stigma of the same flower.

FERTILISATION

Fertilisation follows pollination in which the pollen *male gamete* nucleus fuses with the *female ovule gamete* nucleus forming a *zygote* which becomes the *embryo* of the seed.

Seeds of flowering plants consist of a miniature plant or embryo consisting of a *radicle* and *plumule*, and either *one* or *two* seed leaves called *cotyledons*. The seed is surrounded by the testa or seed coat and fruit wall or *pericarp* (this group of plants are called *angiosperm* or covered-seed plants). *Around* the embryo is the *endosperm*.

GERMINATION

The early growth of the radicle and plumule or germination of the seed will only occur if the following essential conditions are present; *warmth* for the enzymes to change stored food into simple foods, *water* to cause the enzymes to act, and *oxygen* or air to release the energy from food.

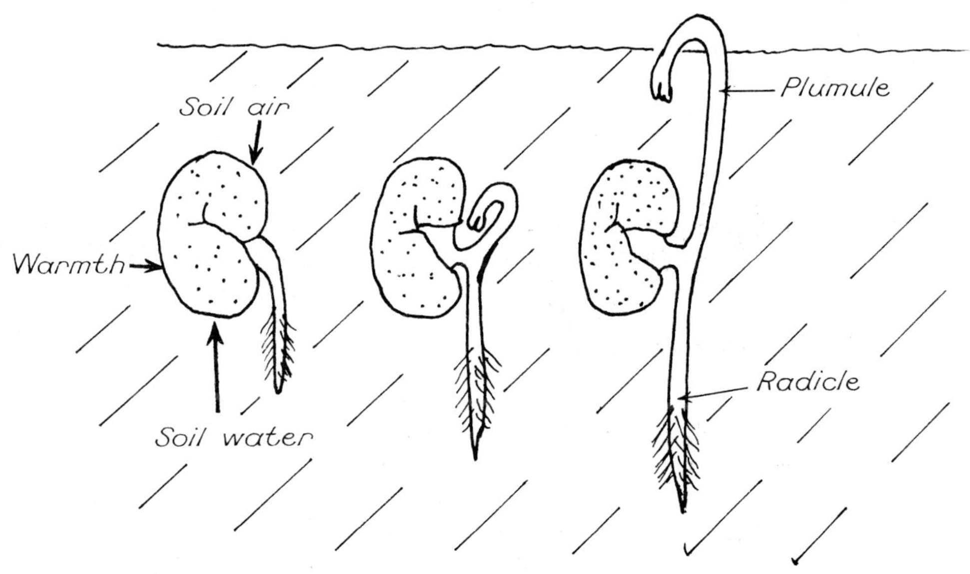

Fig. 6.9 Germinating seed of a flowering plant

MONOCOTYLEDON AND DICOTYLEDON PLANTS

Flowering plants have either one or two seed leaves or cotyledons, this leads to a division of the two main groups of flowering plants, monocotyledon and dicotyledon plants, each having several different features, summarised as follows:

Monocotyledon	**Dicotyledon**
Seed: one cotyledon	Seed: two cotyledons
Parts of the flower in threes: 3 petals, 3 sepals etc	Flower parts in fours or fives, or multiples: five petals, 4 or 5 sepals etc
Multiply by bulbs and other methods including seeds	Reproduce mainly by seeds
Leaves are long and narrow with parallel veins, e.g. grass leaves	Leaves are broad with a network of veins, e.g. lettuce leaves
Examples: daffodils, tulips, grasses, cereals, and bamboo canes	Examples: roses, peas, cabbage, dandelions and daisies

FLOWERING PLANT STRUCTURE

The *vegetative* system is concerned with feeding and growth, the system being divided into *shoot* and *root*.

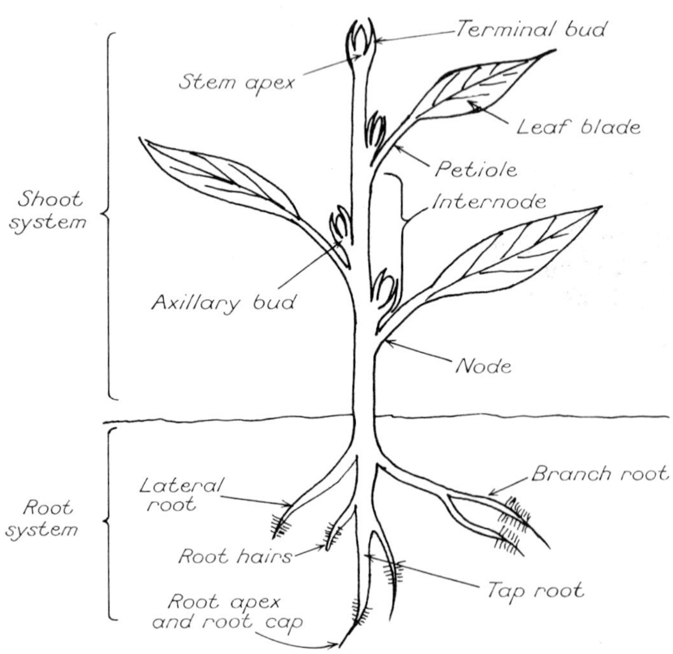

Fig. 6.10 Flowering plant vegetative organs

The *shoot* has a *stem* supporting *leaves* which are attached to the stem at points called the *nodes*. Between the leaf stalk and the stem are found the *axillary* buds, whilst a *terminal* bud, or a flower, is found at the end of the stem. The shoot system of flowering plants is usually branched with several lateral stems.

The *root systems* are composed of a main *tap* root, in dicotyledon plants, together with several *branch* roots and rootlets. Monocotyledon plants have *fibrous* roots without a main tap root. All root systems have tiny *unicellular root hairs* which are essential to absorb water and salts from the soil.

ANNUALS

Plants which grow from seed to produce flowers and seeds, within a fruit, in *one year* are called *annuals*.

BIENNIALS

Plants which take two years, or two seasons, before they are able to flower and form seeds are called *biennials*. Usually the biennials produce a considerable root system in the first year as seen in such biennials as carrots, turnips and swedes.

PERENNIALS

Many plants normally flower each year and produce seeds whilst most of the shoot system *remains* above the ground from year to year. Most of the perennial plants are either low growing, woody *shrubs*, or tall growing, woody *trees*. The shrubs and trees which retain their leaves throughout the year are called *evergreens* compared to those which shed their leaves in autumn which are *deciduous*.

FUNGI: THE NONGREEN PLANTS

All the previously described plants have a green colour due to the pigment *chlorophyll*. The *fungi* are those plants which are lacking in chlorophyll.

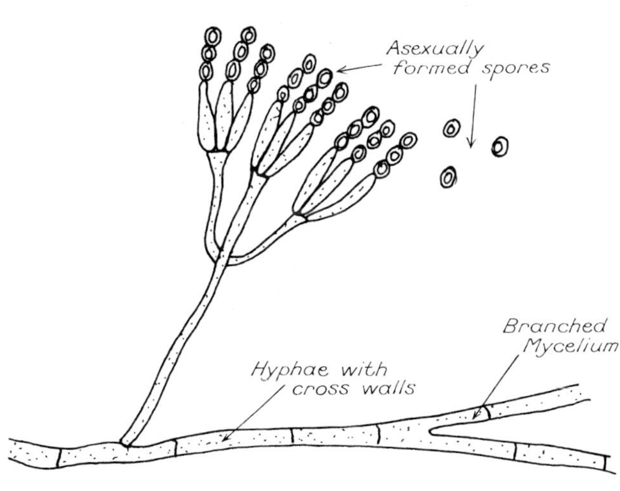

Fig. 6.11 Penicillium mould from which penicillin is produced

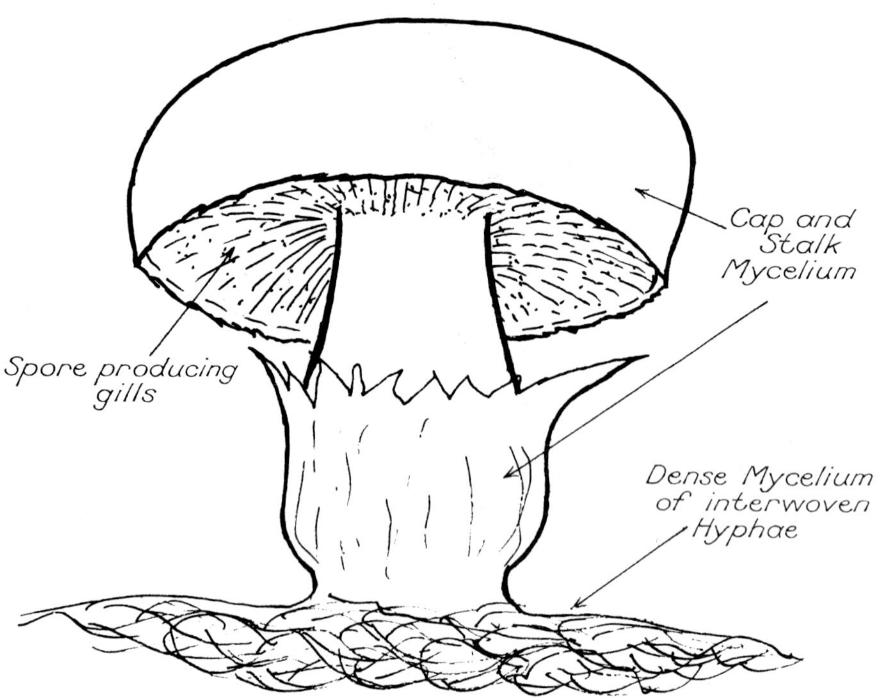

Fig. 6.12 Field mushroom

STRUCTURE

The fungus body ranges from the *single cell* of the *yeasts* to the large multicellular, conspicuous *fruiting bodies*, of mushrooms, toadstools and tree bracket fungi, between these extremes of size are the *threadlike* growths of the moulds.

Apart from the unicellular yeasts, the fungus plant body is composed of branching threads or filaments called *hyphae* which interweave to form a stronger *mycelium*.

REPRODUCTION

Fungi reproduce *asexually* by means of *spores* produced from a *sporangium* rising above the mycelium in most moulds, or which fall from the *gills* of the under surface of mushroom caps. The spores are very small, light, and readily airborne, and will germinate on a suitable food rich medium. Sexual reproduction also occurs in some fungi (Fig. 6.13).

FEEDING

Since fungi are without chlorophyll they are unable to manufacture their own food by photosynthesis. Fungi feed on *complex organic foods* in the living tissues of plants and animals as *parasites*, or on dead and decaying plant and animal material, as *saprophytes*; in the latter fungi play an important part in the process of decay and fermentation.

Fungi secrete the essential *enzymes* needed to digest the complex foods, in addition they *excrete* valuable waste products called *antibiotics*.

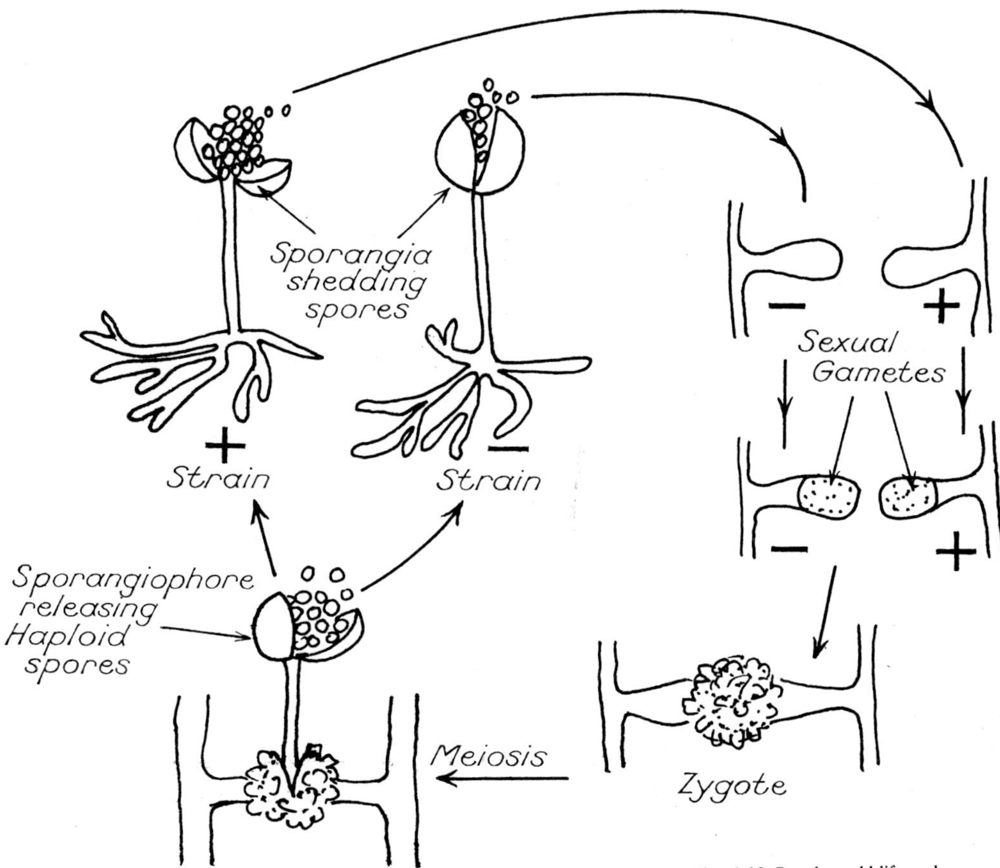

Fig. 6.13 Bread mould life cycle

USEFUL FUNGI

Apart from their ability to feed on plant and animal remains, and act as nature's very efficient refuse disposal corps in converting such waste into valuable *humus*, the fungi have the following uses:

(1) **Antibiotics.** These are used in the treatment of many diseases and are produced as by-products of fungus metabolism. Some named examples include, *penicillin, fumagillin, griseofulvin* and *cephalosporins*.

(2) **Nutrients.** Yeast extracts are valuable sources of amino acids, proteins and vitamin B_2, *riboflavin*, useful in convalescent diets.

(3) **Fermentation products.** Different yeasts are used in the production of wines, spirits, and beers, all of which are fermentation products containing varying amounts of *ethanol* (ethyl alcohol). *Carbon dioxide* is another by-product of the fermentation process and it finds important use as a raising agent for baked foods.

Plate 23 Photomicrograph of a hair infected with a fungal mould causing a mycosis (Bayer Chemicals Ltd)

FUNGI AND DISEASE

Diseases of man caused by fungi are called *mycoses* which are of two main kinds, those attacking the *keratin* of the hair, finger nails and skin are called the *ringworm* disease of the scalp, face, body, feet and nails; and the fungi attacking the *internal* body surfaces such as the mouth, vagina and lungs causing diseases of the mouth, '*thrush*'; of the lung, '*farmers lung*'; and of the vagina as *candidiasis*.

7 The Microbes

Bacteria, together with *unicellular fungi* and protozoa, are of microscopic size and are called *microbes*. Their study is called *microbiology*.

Bacteria are considered, by the majority of biologists, to be *plants* in a unicellular form which are devoid of chlorophyll and may be remotely linked with fungi.

STRUCTURE

The structure of bacteria is best seen through the examination of photographs called *electron micrographs* produced with the powerful *electron microscope*.

Bacteria cell shapes are either *spheres* called *cocci* or *rods* called *bacilli*. In addition these shapes are found in variations as pairs or clusters amongst the cocci, spirals, comma shapes, and flagellated forms amongst bacilli (Fig. 7.1).

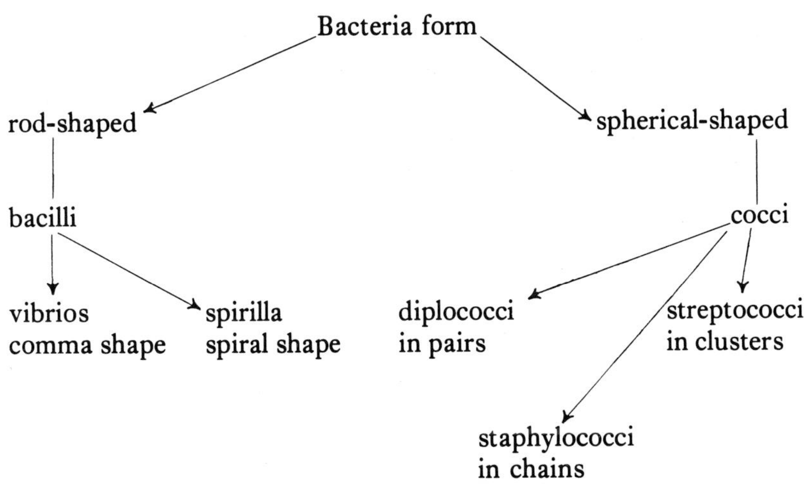

The cell wall of bacteria is fairly *thick* being composed of carbohydrate and protein substance called a *mucopeptide*, in contrast to the *cellulose* of green plants. Around the thick cell wall is a *capsule* or *slime layer*. A *cytoplasmic membrane* is contained within the cell wall having numerous inwarding projecting tubules lined

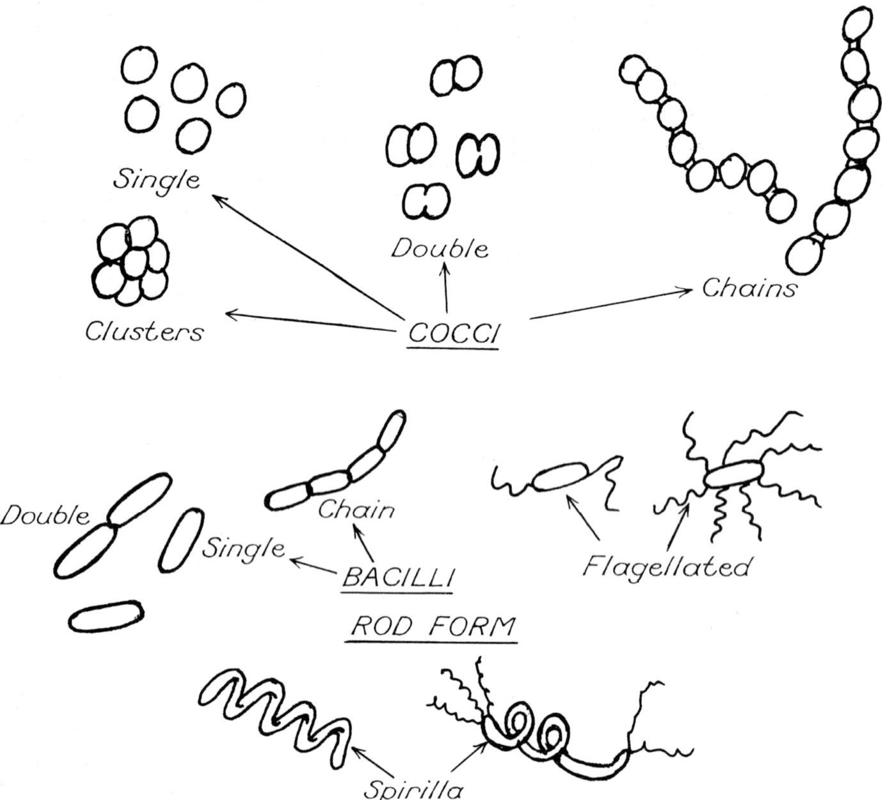

Fig. 7.1 Bacteria cell shapes

with *ribosomes*, similar to those of the *endoplasmic reticulum* of most plant and animal cells (see Fig. 7.2, p. 98).

A distinct nucleus is *not* found in the cell, and the important DNA (deoxyribonucleic acid) material of all nuclei, is found as a ring-shaped structure consisting of only *one chromosome*, this exerts the same influence in the bacteria cell as the larger and complex nucleus of plant and animal cells.

Flagella when present in bacteria cells are found to arise as part of the cytoplasm.

REPRODUCTION

Bacteria reproduce rapidly in soil, water, air or decaying matter. The following conditions are essential to the growth and multiplication of most bacteria in the *human body*:

(1) **Food.** In a simple predigested soluble form, e.g. as provided by serum and plasma fluids of the body. Special laboratory *cultures* are prepared containing the soluble foods from blood plasma or other sources contained in nutrient *agar gels*. In this way bacteria can be grown *outside* their normal habitat of the bodies of plants or animals or soil.

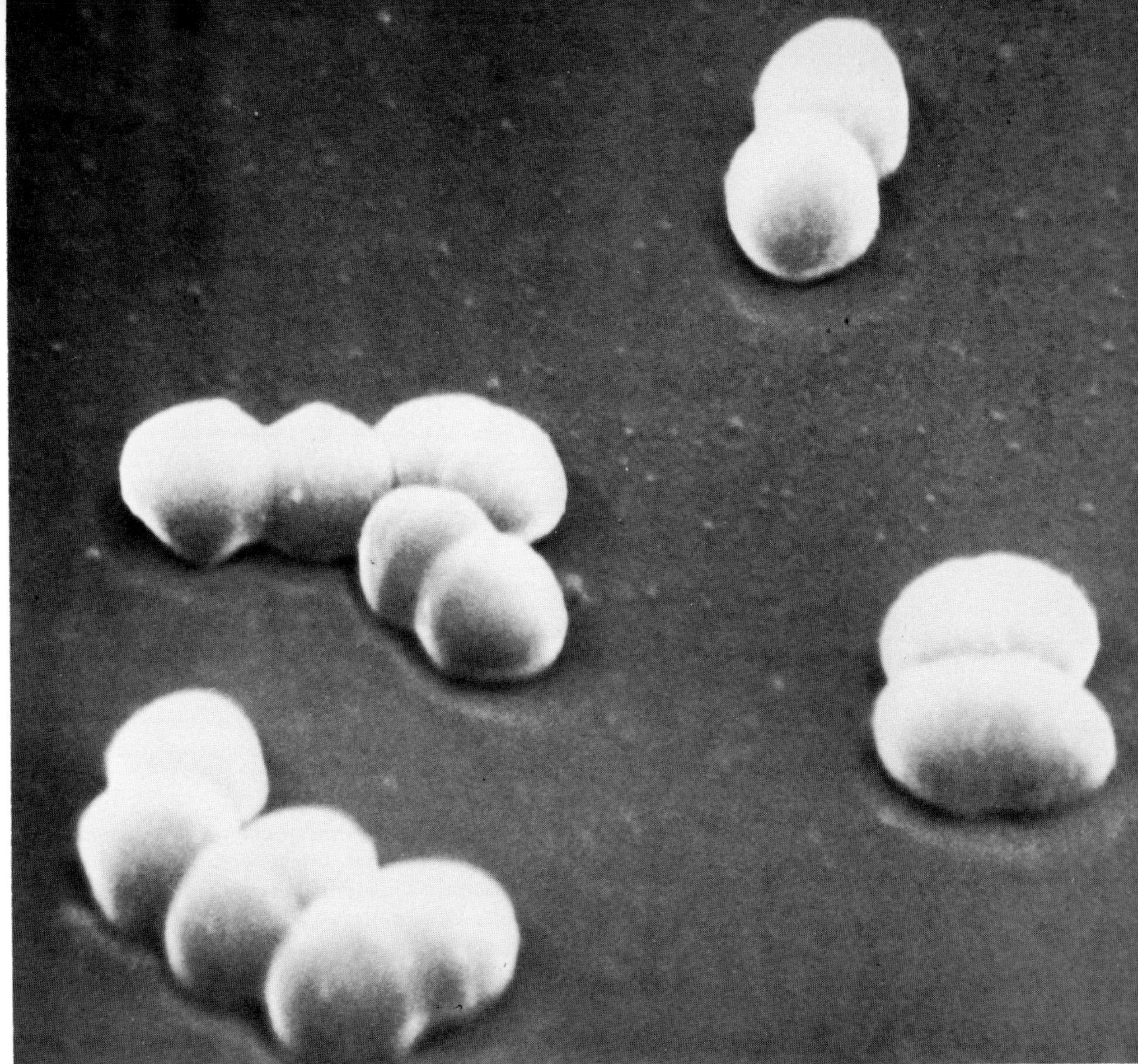

Plate 24 The appearance of the bacteria *Streptococcus viridans* magnified 21 500 times (Upjohn Ltd)

(2) **Warmth** closely similar to normal body temperature or blood heat 38°C is ideal for microbe *incubation.*

(3) **Moisture** is essential.

(4) **pH** should be near to slightly alkaline conditions or neutral.

(5) **Darkness** or lack of sunlight is preferred.

(6) **Oxygen** is not needed by many bacteria, but some bacteria can live in the presence or absence of oxygen. Bacteria needing oxygen are called *aerobes,* whilst those not requiring oxygen are called *anaerobes,* the bacteria which live in either aerobic or anaerobic conditions are called *facultative.*

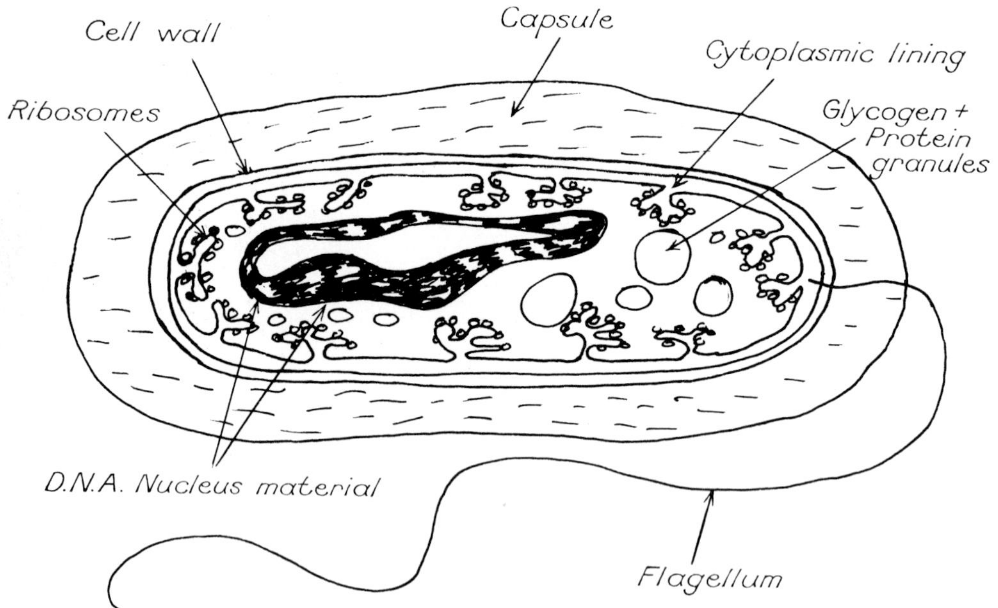

Fig. 7.2 General structure of a bacteria cell

GROWTH AND MULTIPLICATION

In the favourable conditions described on the previous page, which could be provided in the dark, moist, airless, warm, food-rich cavities of the human body, bacteria will multiply by *binary fission* or mitotic division of the nucleus. One bacteria can produce a great number of offspring in the space of 48 hours, often in sufficient numbers to be seen by the naked eye as coloured growths or *colonies* on the surface of a laboratory culture of nutrient agar.

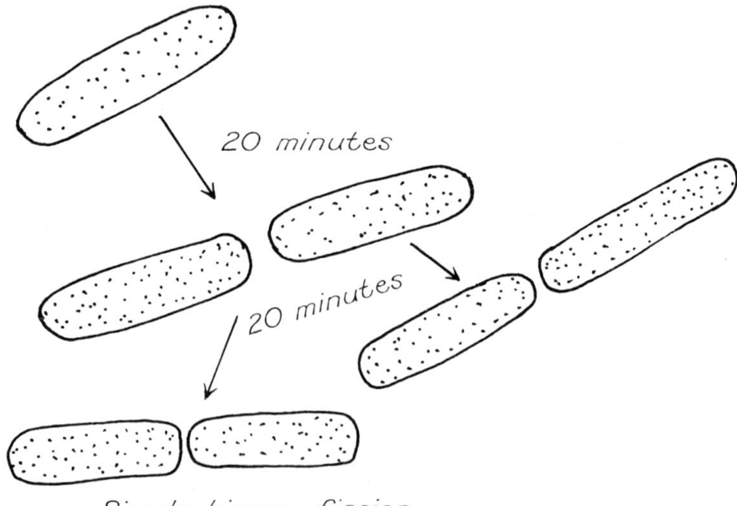

Fig. 7.3 Reproduction of bacteria

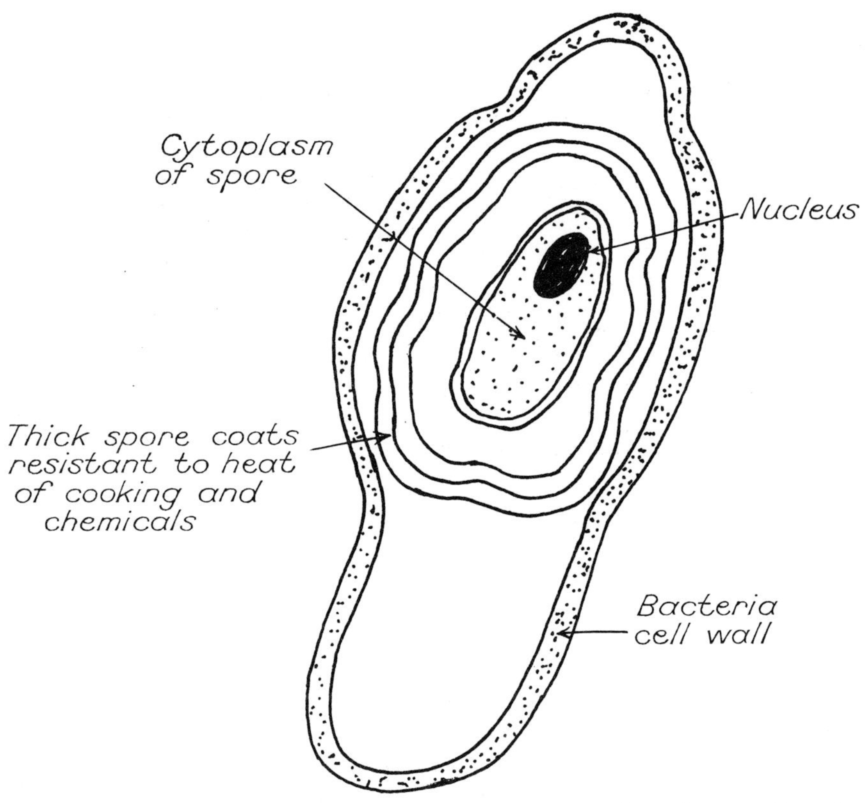

Fig. 7.4 Bacteria spore within a cell

SPORES

Bacteria can surround themselves with a *thick spore coat*, in this form the bacteria can survive for many years in dirt and dust, and also withstand the strong heat in cooking of foods and the effect of certain disinfectant chemicals. Spores therefore enable bacteria to be a dangerous cause of disease since they are hard to destroy.

PATHOGENIC BACTERIA

Pathogens are organisms causing disease in man, many bacteria are pathogenic whilst great numbers of bacteria are also *harmless* to man and are found covering and inhabiting his body in a harmless *commensal relationship* or harmonious living. Others are *symbiotic* being a beneficial relationship, such bacteria are found in the large intestine and can produce vitamins of the B group which are used by the human body.

Harmful bacteria cause disease in man by the *toxins* which they produce as byproducts of metabolism. The toxins *secreted* from the bacteria cell into its surroundings are *exotoxins*, whilst those accumulating within the bacteria cell are *endotoxins*, released when the cell walls are digested on entering the gut of animals.

In man these toxins can destroy blood cells, and can digest the body tissues allowing the microbes to spread as in *abscesses*.

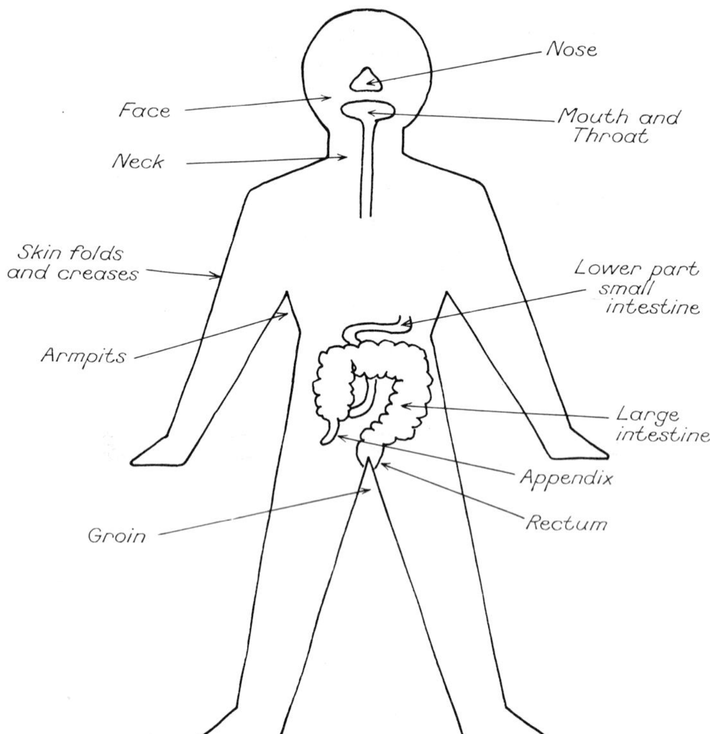

Fig. 7.5 Parts of the human body which harbour bacteria

MICROBES AND MAN

Bacteria are found in great numbers all over the *skin*, in the *mouth* and *throat*, and in the *large intestine* and *rectum* where the *faeces* are composed of great numbers of bacteria.

The remaining parts of a healthy human body are free from microbes, becoming infected with bacteria in disease, or after infection by wounds or postoperatively (Fig. 7.6).

GRAM-POSITIVE AND GRAM-NEGATIVE BACTERIA

A chemical stain called *Gram's stain*, or *methyl violet* is used with *iodine* solution to classify bacteria.

Firstly a *smear* made of the bacterial fluid on a glass microscope slide is dried, or *fixed*. After fixing the Gram stain is added and the bacteria become stained with a violet colour. Treating the stained microbes with ethanol will cause some stained bacteria to *lose* their colour and these are called *Gram-negative*, whilst the others which *retain* their violet stain are *Gram-positive*.

(a) Gram-negative bacteria include: *salmonella* of food poisoning and *coliform* bacteria of faeces.
(b) Gram-positive bacteria are, the *cocci* causing sore throats and boil infections.

In addition to bacteria the fungi, moulds and yeast, can produce a positive result with Gram stain.

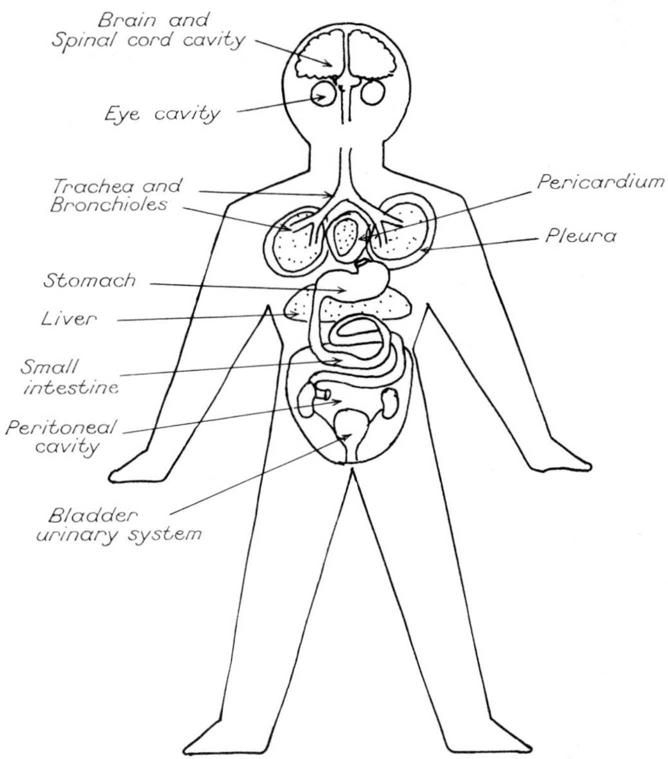

Fig. 7.6 Parts of the human body which are normally sterile

IMPORTANCE OF BACTERIA

(1) **Vitamin production.** Symbiotic bacteria produce vitamins of the B group and vitamin K within the intestine of man. An intake of antibiotics can destroy these beneficial bacteria causing vitamin deficiency disease in man.

(2) **Antibiotics.** More antibiotics are produced by bacteria than by moulds and fungi, such antibiotic producing bacteria are found mainly in the soil and produce: ***streptomycin, neomycin, terramycin***, and the ***tetracyclines.***

(3) **Decay.** The disposal of plant and animal remains as buried refuse, or treated sewage, is due to the action of bacteria and certain fungi and unicellular algae.

BACTERIAL FOOD POISONING

(1) **Infectious poisoning** in which the bacteria, by ***multiplying*** and increasing their numbers in the alimentary canal, cause pain, diarrhoea, vomiting and fever.

(2) **Poisoning by bacterial toxins** made by bacteria which have grown and multiplied on certain foods releasing their ***toxins*** into the foods. The poisoning signs come on rapidly after eating the infected food.

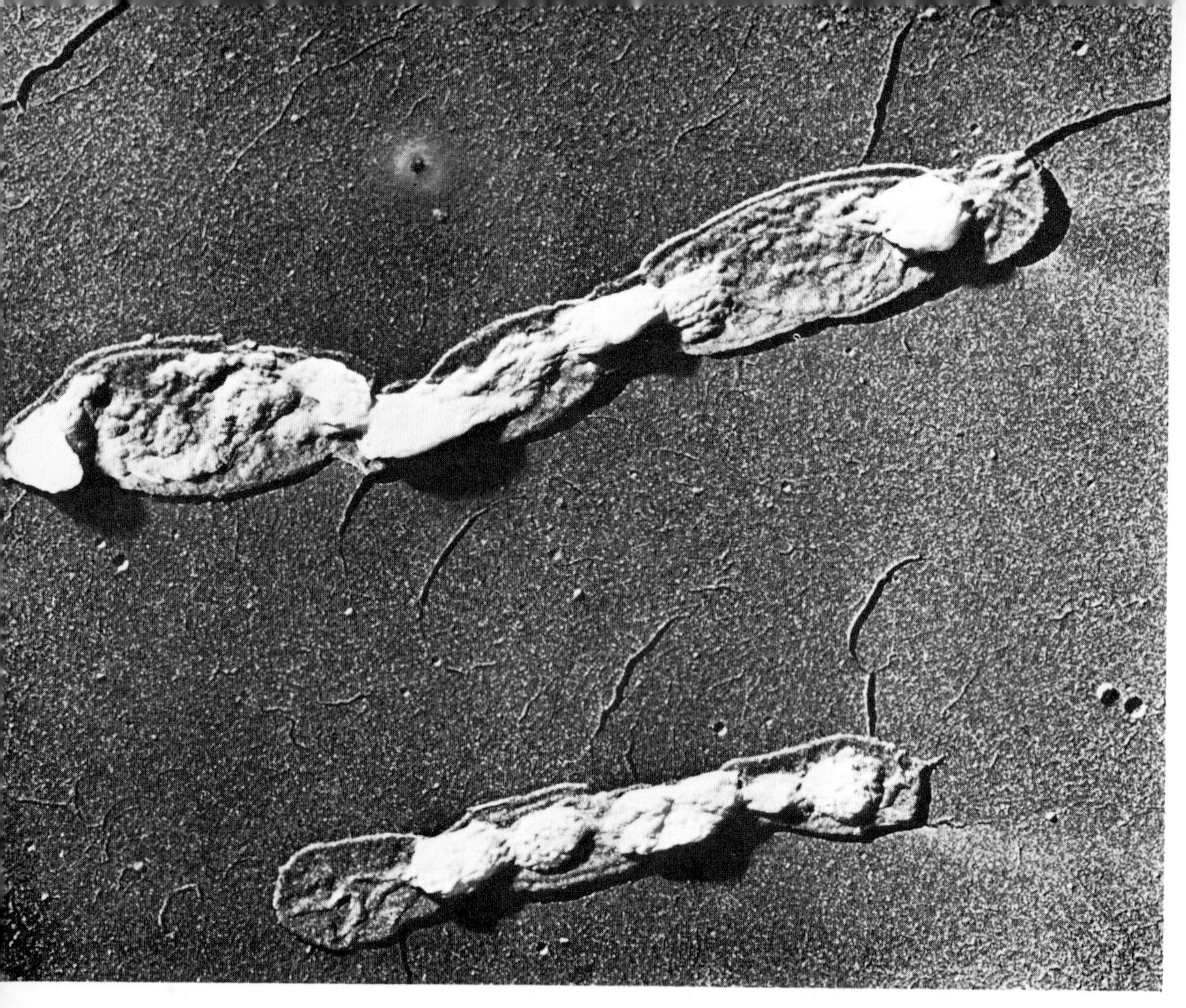

Cooked foods such as meats, stews, milk puddings, creams, custards and gravies are all excellent *culture* media for the growth of microbes, whilst the jellied foods closely resemble the laboratory nutrient cultures.

Infection of the foods, can be by air, water or direct contact by the food handler with unclean fingers, or by animals such as insects, houseflies, cockroaches, rats, mice, cats and dogs. In particular there are infected food handlers, called *carriers*, who may have several pathogenic microbes in their bowel to which they have become immune, and act as a reservoir of infection for non-infected people.

Keeping cooked food in a warm, moist, dark, cupboard will provide excellent conditions to encourage the multiplication of bacteria. Cooked foods should be *rapidly cooled* and stored in refrigerators to prevent microbe growth, unless the food is for immediate use when it should be eaten hot.

Plate 25 Dead cells of a bacteria normally found in the large intestine after attack by the antibiotic ampicillin (Bayer Pharmaceuticals Ltd) (facing page)

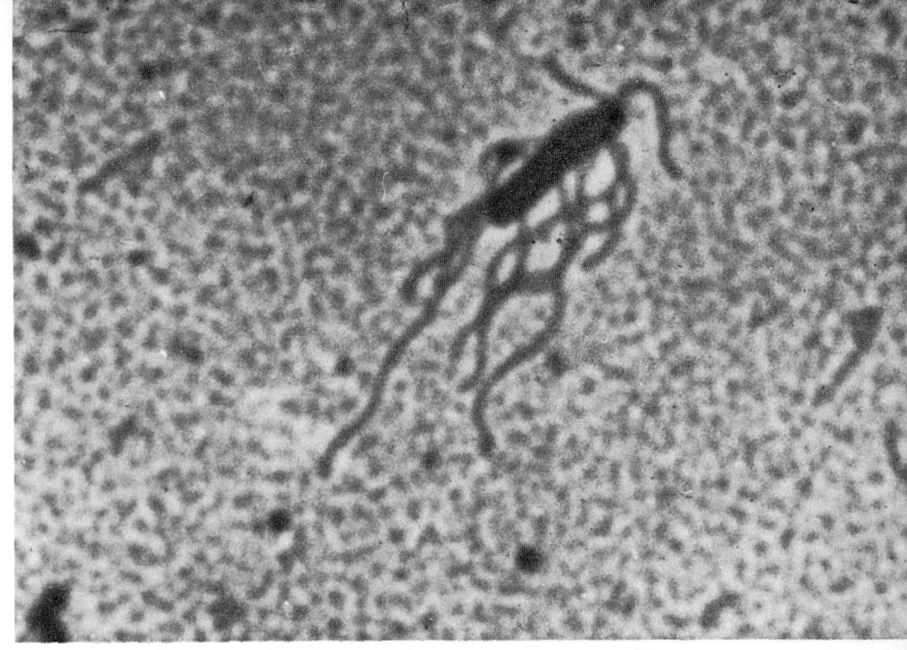

Plate 26 *Salmonella typhimurium,* a bacteria responsible for a form of food poisoning, magnified 6000 times (Unilever Research Laboratories)

FOOD PRESERVATION

Food preservation involves methods aimed at preventing the growth of bacteria, or by methods which destroy bacteria.

(1) **Freezing** slows down the rate of growth of bacteria, but *does not destroy them.*

(2) **Heating** foods aims to make them easier to digest, and also *destroys* most bacteria, except the heat resistant bacterial *spores.*

Heating methods include: pasteurisation, boiling, roasting, grilling, frying, and heating in canning and bottling.

(3) **Dehydration** aims at *removing the essential water* and moisture needed for the multiplication of microbes. Such methods include: the drying of vegetables and fruits by sun, spray, and vacuum drying. Freeze drying is a method in which water is removed from frozen food.

(4) **Chemicals** of various kinds are *permitted* as *preservatives* for use which either, provide an *acid pH* as by vinegar in pickling, or high concentrations of *sugar* in jams and syrups. *Salt* is used in brines for fish and bacon. *Smoking* used for fish and bacon produces certain chemicals which are harmful to microbes. *Sulphur dioxide* is used extensively in fumigating homes and sick rooms following infectious disease. It is also used in preserving fruits and fruit squashes.

(5) **Radiation** with powerful *atomic* rays or *gamma* rays has recently been investigated.

(6) **Antibiotics** are not permitted for use as food preservatives in the United Kingdom.

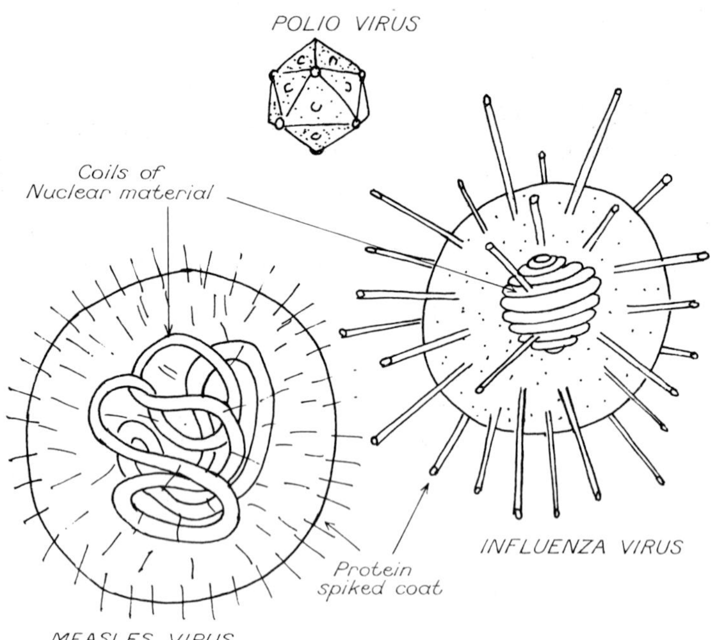

Fig. 7.7 Various forms of viruses

Plate 27 Influenza virus being inoculated into live incubating hens' eggs (World Health Organisation) (facing page)

VIRUSES

Viruses are amongst the smallest living things, being so minute as to be invisible through ordinary light microscope. The study of viruses has been made possible by the *electron microscope.*

Viruses have been called *living proteins*, having the power to multiply themselves, in contrast to many *non-living* protein compounds which are unable to do so. In addition some viruses can exist in *crystal form*, similar to most chemical compounds, these crystals in the presence of living cells can become *alive* and *multiply* themselves. Clearly the viruses occupy a place on the *boundary* between the living and non-living things.

STRUCTURE

The electron microscope shows the viruses to have shapes varying from the simple rods to complex globe shapes, and forms with more than 20 triangular faces called *icosahedrons.*

Nuclear material in the form of a *coil* of DNA (deoxyribonucleic acid) or RNA (ribonucleic acid) is found in the centre of the virus, whilst around it is a coat of *protein* which may be seen as outwardly projecting spikes or prisms that can *attach* the virus to a living plant or animal cell.

NUTRITION

Viruses are only able to feed *within living cells* of plants or animals. Consequently the invasion of living cells by viruses results in cell destruction or damage, evident in many diseases of man caused by viruses. These damaged cells can become attacked by

bacteria in *secondary* infections of human virus diseases. For example influenza virus invades cells of the lung tissues, the damaged cells can be attacked by *bacteria* causing secondary *pneumonia.*

REPRODUCTION

The central part of the virus is either DNA or RNA proteins found in most cell nuclei.

Virus particles entering the living plant or animal cell, cause the host cell to manufacture *new* virus RNA or DNA material which becomes clothed with a protein coat in the form of many new virus particles, these are released by bursting of the host cell. It is the *presence* of the host *cell* nucleus of a live cell, which has the power to *multiply* the virus particles. The consequent breaking out of the virus particles from the cell causes the *damage* which is typical of virus disease in animals or plants. Further infection or *invasion* of more living cells occurs to produce many more virus particles.

Culture of viruses is by means of introducing virus samples to the *living cells* of the developing *chick membranes*, or into the bodies of laboratory *white mice* (Plate 27).

VIRUSES AND MAN

Smallpox, measles, chicken pox, warts, influenza, common cold, mumps and *German measles*, are all examples of some of the commonest virus diseases. These diseases occur *within* the living cells of man; consequently any *drugs* attacking and destroying the virus will *also* attack and destroy the living host cell. In view of this very few drugs are therefore available for treating virus disease.

Virus infected materials such as bedding, clothing, etc., can be sterilised by the usual methods for bacteria which destroy the living organisms by heat, boiling, or the use of chemical disinfectants.

CHANGEABLE VIRUSES

Many viruses, particularly the influenza virus, are able to change their identity in such a way that previous immunity may *not* protect a person from the new *strains* of influenza virus.

A number of viruses have been produced in *crystalline forms* from all or part of the virus. The outer parts of viruses are used to make *crystalline preparations* of the *vaccine* with long keeping qualities.

ROUTES OF INFECTION WITH VIRUSES

Man is infected by disease-causing viruses by the following routes:

(1) **Breathing** in dust and 'droplets' from coughs and sneezes.

(2) **Entry by mouth** in infected food and water.

(3) **Skin** infection through cuts and scratches, or bites from infected dogs (*rabies*), or via many blood-sucking and biting insects.

(4) **Through the placenta** from the mother to the developing baby.

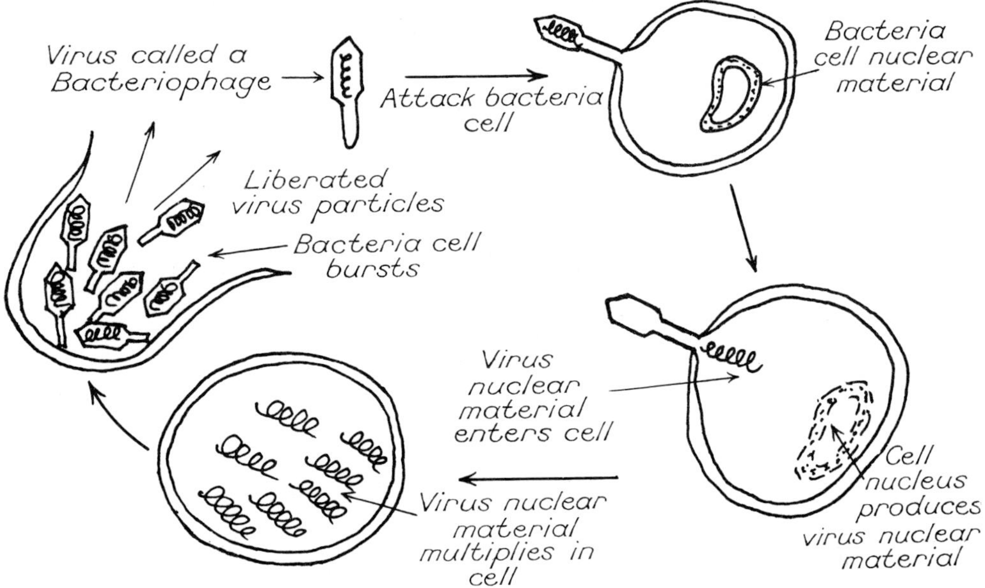

Fig. 7.8 Life cycle of a virus invading a bacteria cell

BACTERIOPHAGES

Certain viruses *can attack bacteria cells* in addition to attacking plant and animal cells; these viruses are called *bacteriophages* or bacteria destroyers. The virus or bacteriophage fixes itself into the bacteria cell and injects its own DNA coil which causes the bacteria cell to make many new DNA coils becoming new bacteriophages that finally burst out of the bacteria cell, and infect further bacteria.

PROTECTION FROM MICROBE ATTACK

Bacteria and viruses which pass through the outer natural barriers of the skin and body interior linings will enter the blood and multiply in the ideal conditions supplied by the blood. Other materials such as proteins, lipoproteins and polysaccharides can be *injected* into the blood. The blood offers certain defence to these substances which act against the body and are called *antigens*. The reaction of the blood to antigens is as follows:

(a) **Antibodies.** Complex chemical substances are made by white blood cells or lymphocytes they are called *antibodies*. A special antibody will act against a certain *antigen*: for example tuberculosis bacteria are inactivated by tuberculosis antibodies, and measles virus are attacked by measles antigen. Surplus antibodies will accumulate in the body. This provides for protection from further attacks of the disease, giving *immunity*.

(b) **Phagocytosis.** The remains of the antigen/antibody combination are engulfed or ingested by large white blood cells by the process of phagocytosis, this leads to the formation of *pus*.

(c) **Anti-toxins.** When the microbes are infecting the body they produce the poisonous *toxins*, the blood cells react by producing *anti-toxins* to neutralise their poisonous effects. The white blood cells produce these anti-toxins, specific ones are needed to combat the toxins of *tetanus* (lock-jaw) or *diphtheria.*

IMMUNIZATION

Natural immunity is seen in man in his resistance to certain diseases which affect other animals, e.g. to foot and mouth disease in cattle.

Active immunity is that which the human body produces by formation of antibodies and anti-toxins, after an infection of the disease or by artificial immunisation through injections of vaccine. The vaccines used in artificial immunisation or vaccination are of the following kinds:

(a) *Living microbes* which are 'near dead' or very aged bacteria or viruses are called *attenuated* microbes. When injected these cause the blood to produce its own stock of antibodies against this disease and may give a lifetime resistance against such diseases as *measles, German measles, tuberculosis, poliomyelitis,* and *rabies.*

(b) *Dead microbes* in sterilised preparations are injected causing the blood to produce antibodies against these antigens, for protection against *whooping cough, influenza, typhoid, cholera,* and *plague.*

(c) *Toxoids* are injections prepared from treated antitoxins, and antibodies made in the bloodstreams of other animals or man, and provide short duration immunity against *diphtheria* and *tetanus* and help to start off active immunity against the disease.

8 *Nutrition of Organisms*

All living organisms are composed of chemical *elements* combined together in chemical *compounds.*

The most abundant elements in plant and animal bodies are the following:

Carbon, hydrogen, oxygen, phosphorus, sulphur and *nitrogen; potassium, magnesium, calcium, iron, aluminium* and *sodium.*

These chemical elements enter living organisms as components of food, water or air.

(1) **Water** is a compound of *hydrogen* and *oxygen,* and is essential for all plants and animals.

(2) **Air** is mainly a mixture of *nitrogen* (79%), *oxygen* (20%) and small amounts of *carbon dioxide* and *rare gases.*

(3) **Foods** are complex chemical compounds of the following main groups:

(a) *Carbohydrates,* including sugars, starches and cellulose composed of *varying* amounts of carbon, hydrogen and oxygen.

(b) *Lipids,* as oils and fats, composed of carbon, hydrogen and oxygen; they possess a higher proportion of carbon than carbohydrates, making them 'carbon' rich compounds similar to fuels.

(c) *Proteins,* the main component compounds of *meat* and the cell *nucleus,* formed from a number of chemical elements—carbon, hydrogen, oxygen, nitrogen, sulphur and sometimes phosphorus.

(d) *Vitamins,* present in most foods as complex compounds of varying composition.

(e) *Minerals,* found mainly as *salts* or simpler chemical compounds. For example sodium chloride or common salt is composed of the elements sodium and chlorine.

PLANTS AS FOOD PRODUCERS: PHOTOSYNTHESIS

Animals depend on green plants for their food supply. Green plants are able to manufacture complex foods from simple raw materials, *water* and *carbon dioxide*, using *light energy* and the *catalyst chlorophyll.*

Chlorophyll is a mixture of colour pigments located in cell organelles called *chloroplasts*; these are found all over the plant body or thallus of nonflowering plants, whereas they are mainly in the leaves and stems of flowering plants. Roots and rhizoids are usually without chloroplasts.

Light energy from sunlight, or artificial light, is taken up by the chloroplasts and is made available as *chemical* energy in the form of *adenosine triphosphate (ATP)*, the energy carrying molecule.

Water is split by means of the ATP energy to form *hydrogen* and *oxygen*; the oxygen not needed for respiration escapes into the air as a by-product, whilst the hydrogen attaches itself to the carbon dioxide to produce a chemical compound which becomes the forerunner of *glucose.* On reaching the storage organs, usually located in the plant root system, the glucose is changed into starch.

SUMMARY

(1) Solar or light energy → chlorophyll → ATP

(2) ATP provides energy to split water

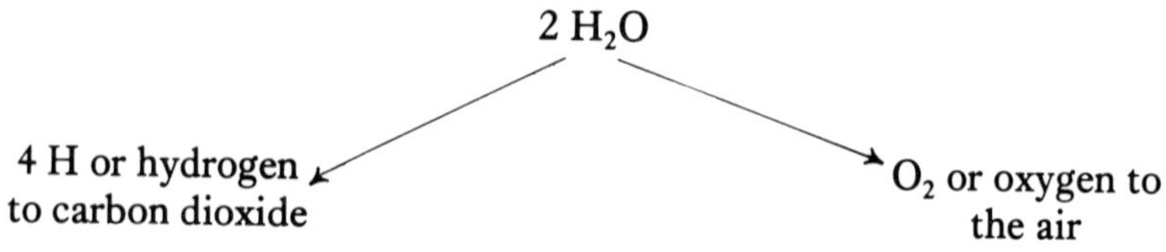

(3) The carbon dioxide is reduced by the hydrogen released from water to produce an *intermediate* compound which becomes *glucose*, and then *starch.*

The complete process of manufacture of glucose or starch, which makes up carbohydrate foods, is called *photosynthesis.* The following chemical *equation* summarises the overall process of photosynthesis:

$$6\ CO_2 + 6\ H_2O + \text{light energy} = C_6H_{12}O_6 + 6\ O_2$$

chlorophyll

carbon dioxide — water — glucose — oxygen to air

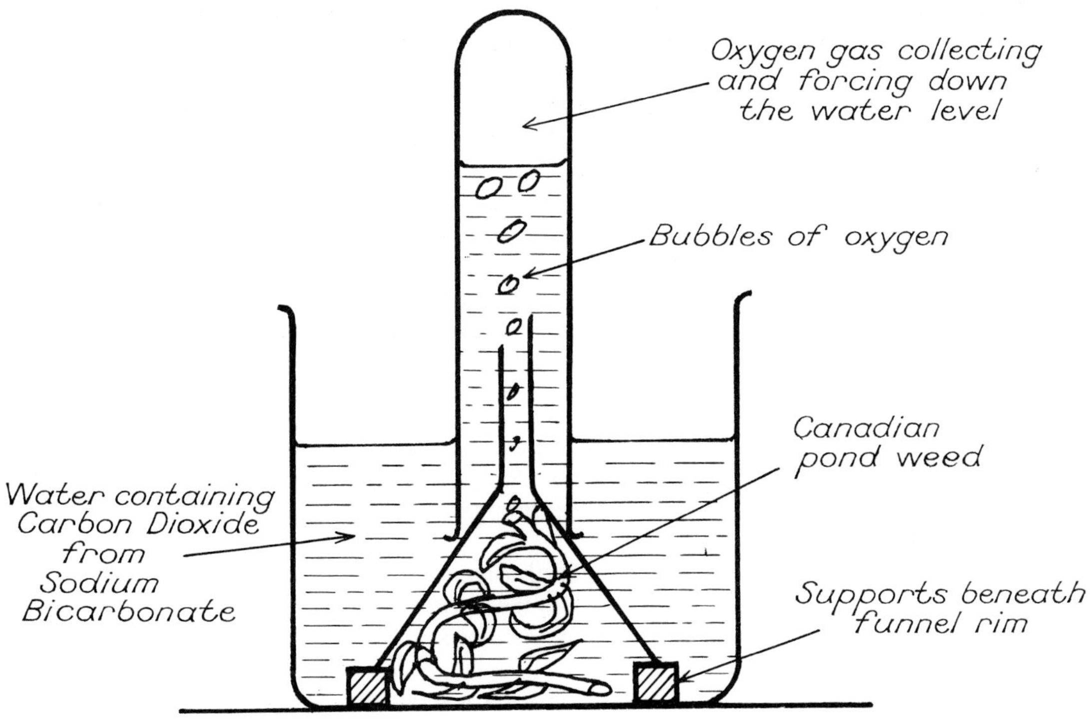

Fig. 8.1 Oxygen is a by-product of photosynthesis

Conditions needed for photosynthesis in plants

(1) Chlorophyll is essential: fungi are deficient in it and cannot synthesise their own foods.

(2) Sunlight or artificial light.

(3) Raw materials, mainly water and carbon dioxide.

(4) Mineral elements such as magnesium and iron are needed to form the chlorophyll molecules.

PRODUCTS OF PHOTOSYNTHESIS

(1) **Oxygen** enriches the air or surrounding water. Algae are used to produce oxygen in order to destroy harmful bacteria in water purification.

The production of oxygen by water plants can be shown in the following experiment: a piece of pondweed or the algae *spirogyra* is placed in an inverted funnel with a test tube full of water connected to the funnel stem. The apparatus is left exposed to light, and the accumulation of oxygen bubbles in the test tube is noted. When sufficient gas has collected in the test tube it may be tested by means of a glowing splint.

(2) **Starch** is rapidly formed from glucose. It may be tested for in a green leaf as follows: take a *variegated leaf* (one with green and yellow patches), place it in boiling water and then transfer it to a tube of ethanol which is heated to boiling point in a water bath, to remove the chlorophyll pigments. Dip the leaf in boiling water once more. Transfer it to a solution of iodine in potassium iodide solution; it will develop blue-black coloured patches where starch is present whilst the original yellow or non-green parts will not develop the blue-black colour.

AUTOTROPHIC NUTRITION

All green plants manufacture food by photosynthesis and the method of feeding is called *autotrophic*. Some bacteria are autotrophic in using *chemical* energy for their food making process.

ANIMALS AS FOOD CONSUMERS

Food chains show the dependency of animals on plants as sources of food and in turn show how certain animals feed on other animals. The following food chain shows how the energy flows from the sun through plants to man:

(a) **Phytoplankton** are small green algae which photosynthesise their food in freshwater and seawater; similarly the *grasses* undergo photosynthesis on land.

(b) **Zooplankton** are small embryos and tiny shellfish feed as herbivores on the phytoplankton.

(c) **Fish** as gillfeeders, e.g. herring, feed upon both phytoplankton and zooplankton, whilst *herbivorous* cattle feed on grass.

(d) **Carnivorous** fish, e.g. ray and skate, and carnivorous animals, e.g. the fox, mainly feed on herbivorous animals.

(e) **Omnivorous** man can eat either plants or animals, thus deriving energy from the sun which has passed through the food chain.

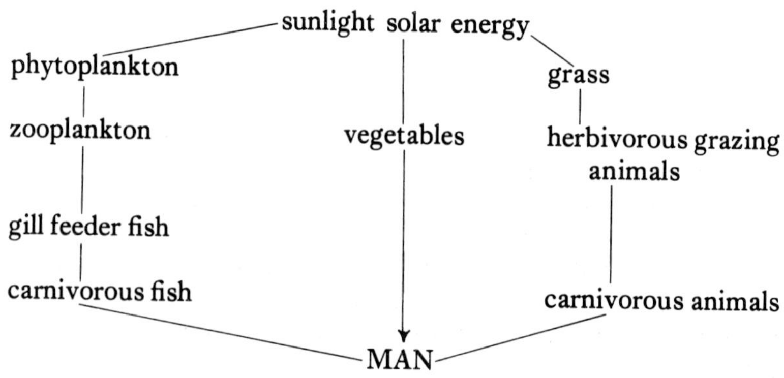

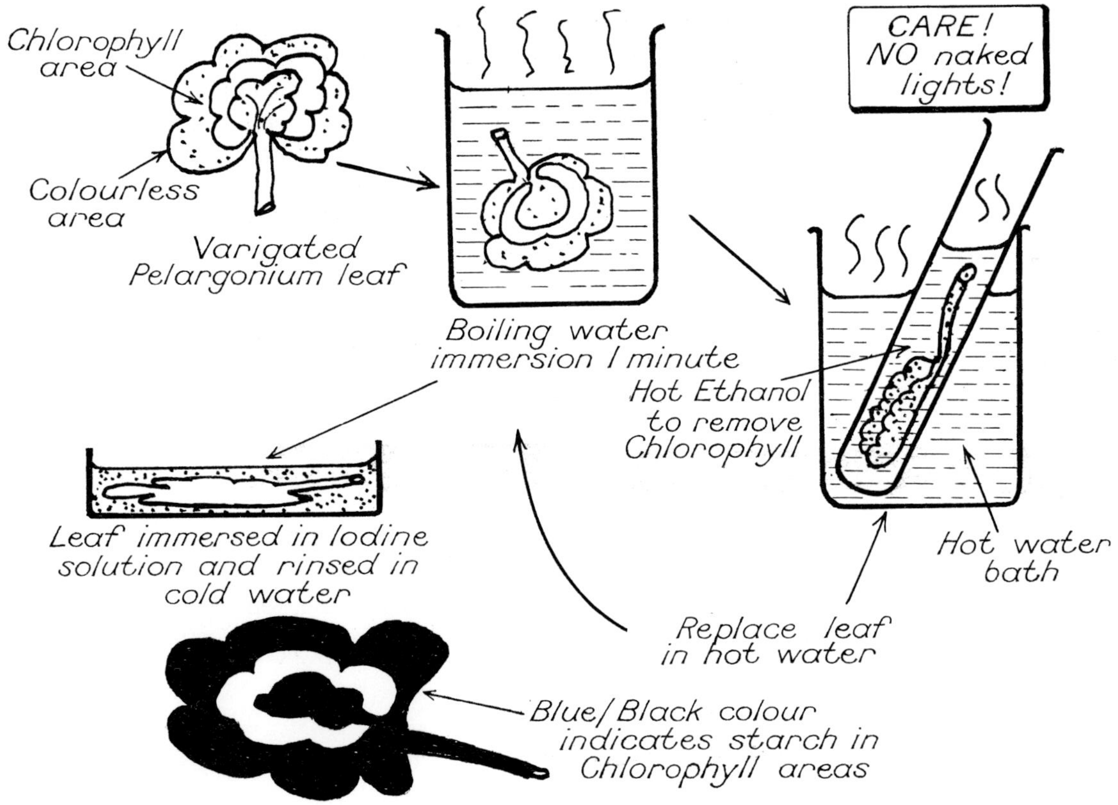

Fig. 8.2 Starch is produced by chlorophyll in a variegated leaf

FOOD PYRAMIDS

These show the great ***numbers***, or ***weight*** of organisms, needed to support *one* man. Millions of tons of plankton feed the large numbers of fish eaten by a man in his lifetime. An estimated 1000 kg of plankton forms 1 kg of flesh in a man.

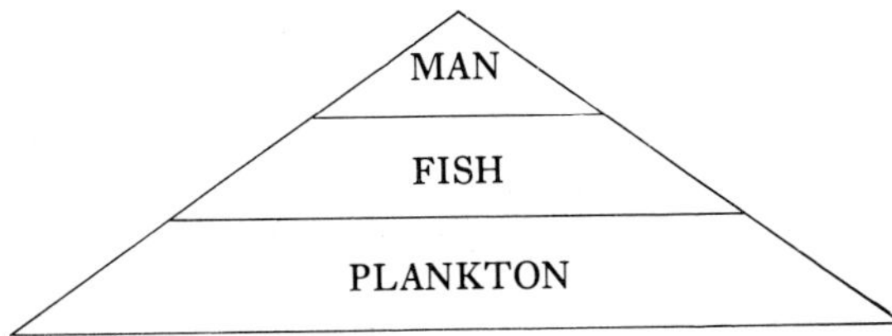

FOOD WEBS

Food webs are the *complex* feeding patterns formed by interconnecting many different food chains to support life in a natural community. If a *single* chain is broken, organisms usually adapt to another.

HETEROTROPHIC NUTRITION

This is the method of feeding requiring a supply of *organic foods*, carbohydrates, proteins and lipids, produced by autotrophic plants. Herbivores, carnivores, most bacteria, and fungi are heterotrophic organisms.

INGESTION OR ANIMAL FEEDING METHODS

The main methods of ingestion in animals are:

(1) **Phagocytosis** or the engulfment of food particles by the animal cell is seen in the protozoa, and white blood cells of man. *Pinocytosis* occurs as a process of ingestion by means of cell microvilli see pp. 9 and 117.

(2) **Suction** of food in a liquid form: e.g. milk is sucked into the mouth of young mammals, or drawn up a *proboscis* in the mouthparts of the housefly, mosquito, lice and fleas.

(3) **Filter feeding** is the drawing of water containing food through a filter formed either by the *gills,* seen in fish and certain molluscs, or through a comb-like extension of the jaw called the *baleen* in whalebone whales.

(4) **Tongue** feeding is used by amphibia, reptiles and the anteater mammals. The long coiled tongue flicks out and gathers the mainly insect form of foods.

(5) **Teeth** are found amongst the vertebrate animals.

Most vertebrate animals use their teeth to *hold* the prey or food by means of sharp pointed conelike teeth similar to *canine* teeth of man. Carnivorous mammals have sharp molar teeth which are pointed and suited for holding, stabbing and tearing the food, followed by a limited form of shearing or cutting before swallowing.

Herbivorous mammals have teeth with toughened flat furrowed crowns passing through each other during prolonged grinding by the specialised *molar* teeth; in addition there are the sharp cutting *incisors.*

Omnivorous animals, including man, have teeth intermediate between the herbivorous and carnivorous animals, composed of sharp cutting *incisors,* pointed *canines,* and partly pointed and cusped *premolars* and *molars,* for cutting, crushing and grinding food.

Birds which are without teeth grind the food with the aid of swallowed stones in a muscular gizzard.

(6) **Intravascular feeding.** Soluble digested foods may be passed directly into the blood by intravascular or *intravenous infusion* when a vein is pierced to introduce the food material in a sterile solution.

The embryo mammal receives food into its blood capillary network in the *placenta* from the blood capillary network of the mother. Foods such as glucose, amino acids, fatty (alkanoic) acids and glycerine (propanetriol) pass by *diffusion* from the mother's blood into the embryo's blood. Fluid surrounds the embryo blood vessels allowing *intervascular* feeding. The process resembles *haemodialysis*.

DIGESTION

Digestion is a process in which complex *insoluble* foods are changed into simple *soluble* foods by a chemical reaction called *hydrolysis* in the presence of biological catalysts called *enzymes*.

The large molecule compounds of carbohydrates, lipids, and proteins are split by the chemical action of water this chemical change is speeded up by the enzyme catalyst, which is best able to function at blood heat 38°C.

Unlike *chemical* catalysts, the enzymes can only control a *specific* change, for example the lipids can only be hydrolysed by lipid enzymes called *lipases*.

The specific enzyme name always ends in *-ase*.

Carbohydrates, lipids and proteins produce the following end-products of digestion:

(a) Carbohydrates → GLYCOSIDASES → glucose type, monosaccharide
(b) Lipids → LIPASES → fatty acids (alkanoic acids) and glycerol (propanetriol)
(c) Proteins → PROTEASES → amino acids

Intravenous infusions will consist of specially prepared *sterile* solutions of mixtures of monosaccharides, amino acids, fatty (alkanoic) acids, glycerol (propanetriol) together with added vitamins.

CONDITIONS FOR FOOD DIGESTION

(1) Blood heat 38°C
(2) Water
(3) Specific enzyme
(4) Certain pH. Changes in the pH in different parts of the alimentary canal of man are seen as follows: in the mouth pH 6.8, stomach pH 1.0, small intestine pH 8.4, and rectum pH 8.0. The strong acid conditions of the stomach is necessary to activate the enzymes and also helps to destroy certain harmful microbes.

Bacteria are microbes which play an important part in the digestion of *cellulose* in herbivorous animals, since the alimentary canal cannot produce digestive enzymes for this material which forms the roughage of a human diet.

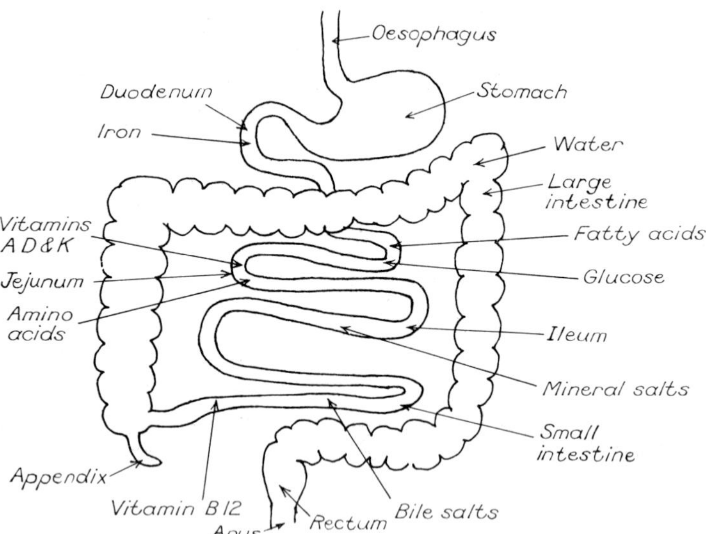

Fig. 8.3 Food components and the position of their absorption in the alimentary canal

ABSORPTION

The products of digestion (glucose, fatty (alkanoic) acids, glycerol (propanetriol), and amino acids, vitamins and minerals) are absorbed into the small intestine of man at certain sites shown in Fig. 8.3.

Surgical removal of parts of the intestine will result in shortages of different nutrients in the body; this condition, called ***malabsorption***, may in turn cause deficiency disorders of the skin, bone and blood.

Villi, which are finger-like projections of the small intestine wall, help to increase its total internal surface area. A similar structure, the ***typhlosole***, is seen as an infolding of the earthworm intestine to increase its surface area for purposes of absorption.

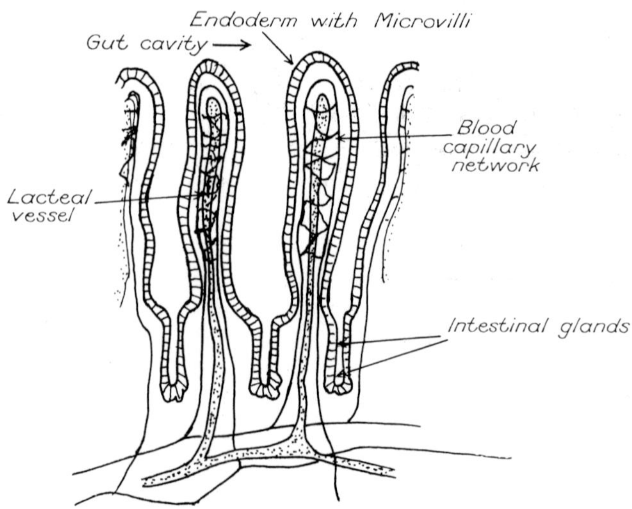

Fig. 8.4 Structure of small intestine villi

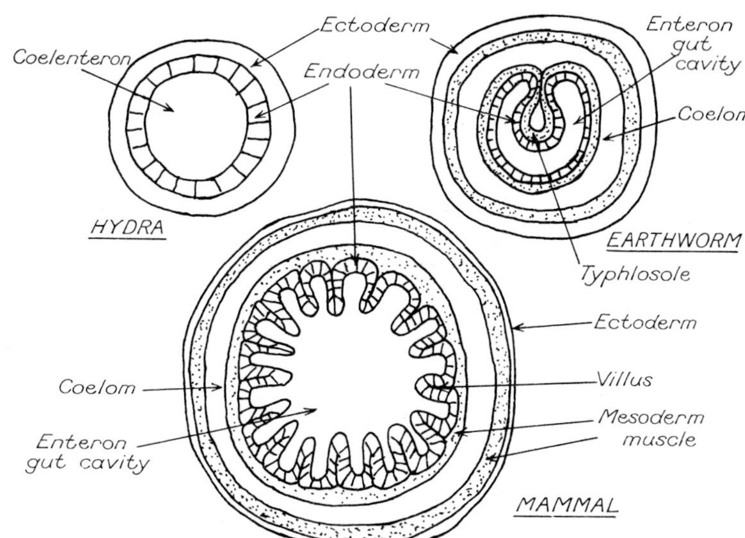

Fig. 8.5 Methods of increasing the intestine absorption surface

The epithelium lining the intestine of man has ***microvilli*** on the surface facing the gut cavity; these are in effect miniature villi and are able to absorb the digested food and pass it into the capillary blood system or lacteal vessel. Glucose and amino acids pass by ***diffusion*** into the blood capillaries, whilst fatty (alkanoic) acids and glycerine (propanetriol) diffuse into the lacteal vessel.

Pinocytosis is a process of ***engulfment*** of fluid and solid food particles and occurs between the microvilli of the small intestine. The pinocytes are small vesicles that pinch off the food in droplets.

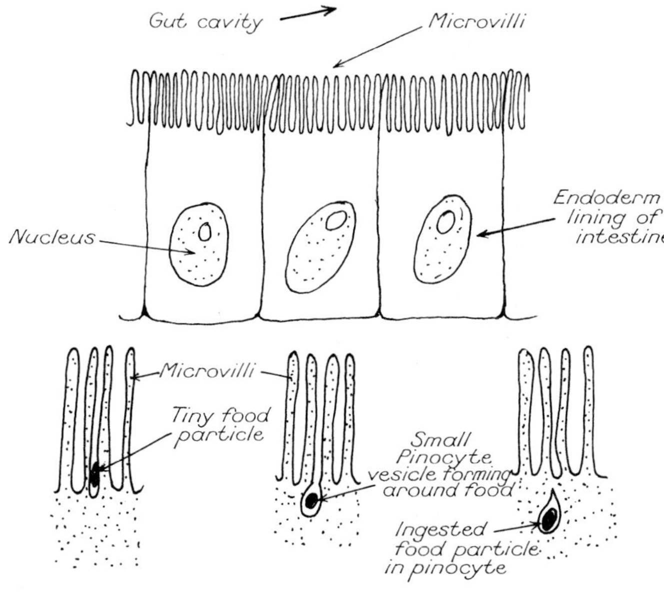

Fig. 8.6 Microvilli and pinocytosis

PARASITISM

In previous chapters reference has been made to animals and plants as parasites. *Parasitism* is a relationship between two living organisms in which one lives at the expense of the other. The parasite derives benefit from the *host*, whilst the host often suffers harm from the relationship.

External parasites of the animal host body are called *ectoparasites*, whilst the internal parasites are called *endoparasites*. The following is a summary of the parasites on man:

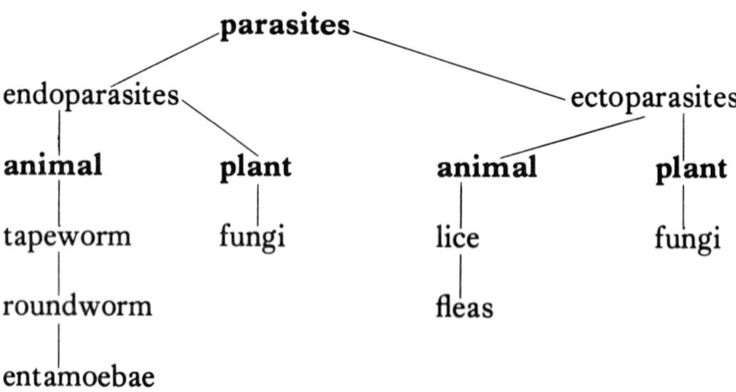

MUTUALISM OR SYMBIOSIS

Mutualism is a beneficial relationship of two or more living organisms. Mutualism is seen in the large intestine of man, where certain bacteria produce vitamins of the B group in return for food and shelter. *Lichens* are a mutual relationship between a *fungus* and *algae* which inhabit its mycelium. The food manufactured by photosynthesis of the algae is shared together with the oxygen; the fungus mycelium provides shelter and a supply of water and salts in return.

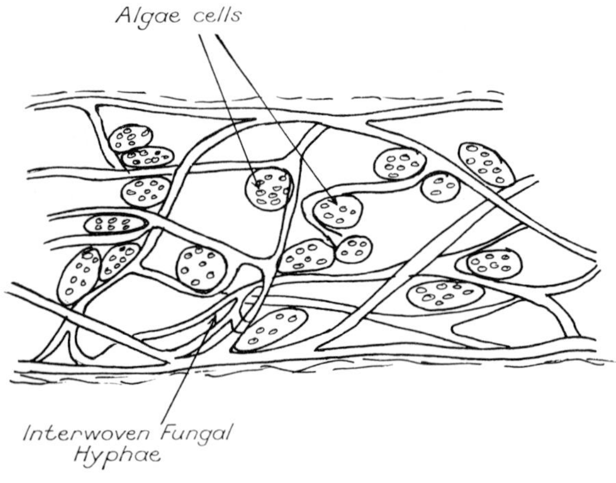

Fig. 8.7 Structure of a lichen

SAPROPHYTES

Saprophytes are mainly fungi and bacteria feeding on *dead* and *decaying* material of plant or animal origin which break down the substance into a valuable soil component called *humus*.

SCAVENGERS

Plant and animal remains including the refuse of modern man, is disposed of by the activities of *scavengers* such as dung beetles, rats, mice, vultures, carrion crows, and gulls. They ingest this waste and their faeces become beneficial to the soil.

CYCLING OF CARBON AND NITROGEN

The amount of *chemical elements* on the earth has remained the same since the time of the earth's creation; very little has left the earth except through space exploration.

Many chemical elements enter the body of the plant and animal as food raw material, or as manufactured food, and in turn become important *body building* components. The chemical elements return to the soil, water or air during the life or on the death of the living organism. Hence a *recycling* of the chemical elements takes place continuously between living organisms and the soil, water or air.

An important *link* in the efficient cycling of these chemical elements are the moulds and bacteria responsible for *decay*.

CARBON CYCLE

Photosynthesis requires *carbon dioxide*, changed by hydrogen and solar energy into *carbohydrates*; this in turn provides energy as a food for plants and animals. The product of respiration of food in living organisms is *carbon dioxide*.

The *decay* of plant and animal remains yields more carbon dioxide through the activity of moulds and bacteria.

Fuels, such as coal, wood, oil and gas, are called 'fossil fuels' the product of plant and animal decay; all these fuels have a high *carbon* content which on oxidation during 'burning' releases carbon dioxide as one of the products of combustion.

Limestone, or calcium carbonate, formed from the remains of shellfish, will 'lock up' some carbon dioxide; thus some carbon is unable to participate in the cycle.

```
                        photosynthesis
GREEN PLANTS ──────────────────────────→ CARBOHYDRATES
     ↑                                        │
CARBON DIOXIDE                                │
     ↑                                        ↓
     ├──RESPIRATION ←──────────────────┬─PLANTS
     │                                 │   ↓
     ├──DECAY ←────────────────────────┤
     │                                 └─ANIMALS
     └──FUELS ←────────────────────────
```

NITROGEN CYCLE

Nitrogen is another chemical element recycled in a similar way to carbon. Nitrogen passes into the plant and animal body mainly in protein foods which are used for body building purposes. On the death and decay of the living organism the nitrogen is recycled into seawater, soil water, or air. Bacteria and moulds again play an important part in this recycling process.

Nitrogen taken into the plant is manufactured into *protein*, which in turn may become the food of animals or enter the soil on the death of the plant or animal body. The remains of the organisms undergo *decay* first by the action of the *saprophytic* moulds which produce the *humus*. Then *nitrifying* bacteria functioning in well aerated soils change the humus into *ammonia* or *nitrate* compounds. Plants can take in the nitrate salts and change them into *protein* as part of the process of photosynthesis.

Peas and beans are plants which can use *nitrogen gas* present in the soil air; this is done by *nitrogen fixing* bacteria which live in nodules of the plant roots. Thunder and lightning storms will add *nitrates* in small amounts in the rainwater, but more nitrate can be added by artificial nitrate *fertilisers*.

Denitrifying bacteria work in waterlogged soils and in poorly aerated soils, causing the nitrates to be changed into nitrogen gas escaping into the air.

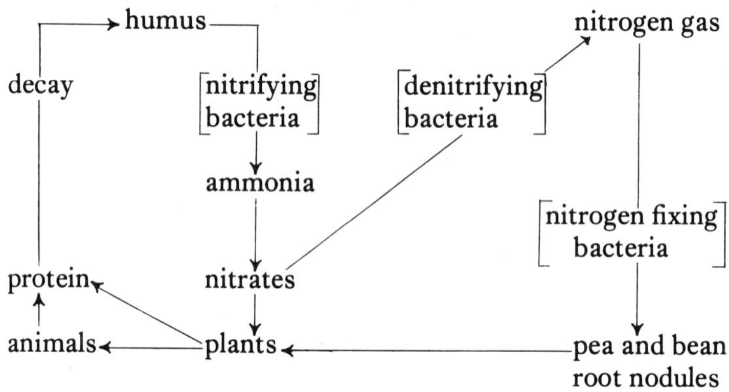

SOIL

Soil is essential for plant and animal life on land and its *composition* is shown to be a *mixture* of the following:

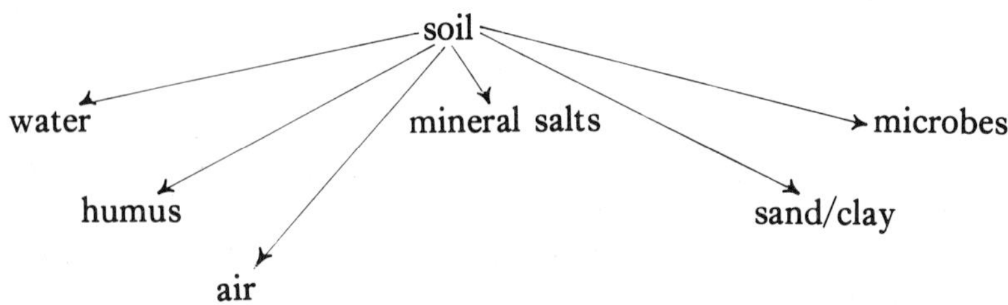

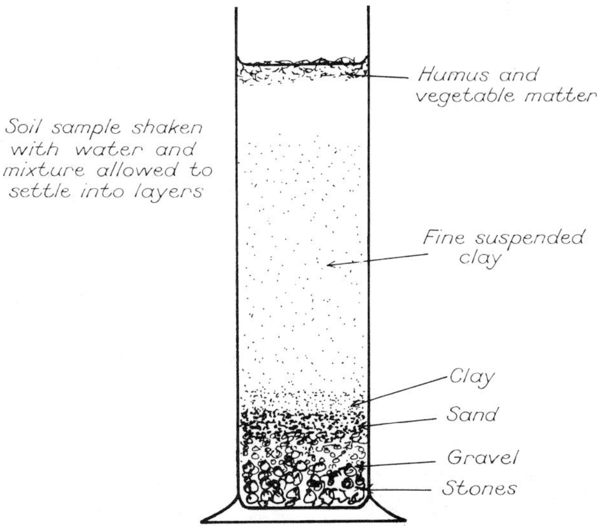

Fig. 8.8 Soil analysis

Water is essential to support the life of plants and soil dwelling inhabitants.

Humus, apart from being a source of nitrogen for plants and microbes, serves also to hold water like a sponge, and bind the soil particles tovether in a crumb structure. Soil without humus can blow away as a fine dust.

Mineral salts are present either from the natural component from weathered rock, or as added fertilisers. The main chemical elements needed by growing plants are *nitrogen*, *phosphorus*, and *potassium*. In addition, growing plants need smaller amounts of sulphur, calcium, magnesium and iron.

Sand and clay form the main bulk of soils. They are the weathered and powdered remains of rocks. Sand particles are much bigger than clay particles.

Soil air is introduced during *cultivation* either by digging or other methods which bring air into close contact with the soil, or by the activities of burrowing animals—particularly earthworms. Air is essential for the nitrogen fixing and nitrifying bacteria and other organisms.

Waterlogged soils are acid, due to lack of soil air; such soils are seen in peat bogs where human and animal remains may be found in a comparatively well preserved condition. Sandy soils allow rapid decay and decomposition. This is evident when animal and human remains are found in the sand of beaches and dunes, usually as dry bones.

Seawater

Seawater is important to plant and animal life of the sea and oceans, just as soil is necessary to support life on land.

Seawater is a mixture of dissolved salts and gases. Dissolved salts amount to about 3.6%, made up by the following chemical elements: sodium, magnesium, calcium, and potassium, with the acid radicals—chloride, sulphate, bromide, and iodide. The main compound in seawater is sodium chloride, which is present to the extent of 2.6%.

Dissolved oxygen is essential for respiration of marine life and is present to the extent of 0.8%, whilst carbon dioxide is often present as sodium hydrogen carbonate.

Blood plasma contains 0.9% of dissolved sodium chloride compared to the 2.6% present in seawater.

CIRCULATION AND TRANSPORTATION

If a unicellular plant or animal were surrounded by a *small* amount of water, the oxygen and food contained in this *extracellular fluid* would soon be used up; similarly soluble excretory waste products collecting in this fluid, could become *toxic* to the cell. *Heat* liberated during respiration could cause the extracellular fluid to *overheat* the cell. Most unicellular plants and animals are however surrounded by *large amounts* of either freshwater or seawater which rapidly remove excretory waste and surplus heat.

Multicellular organisms, other than coelenterates and certain plants, have cells surrounded by smaller amounts of extracellular fluid. These are linked to a *circulatory* system which brings the

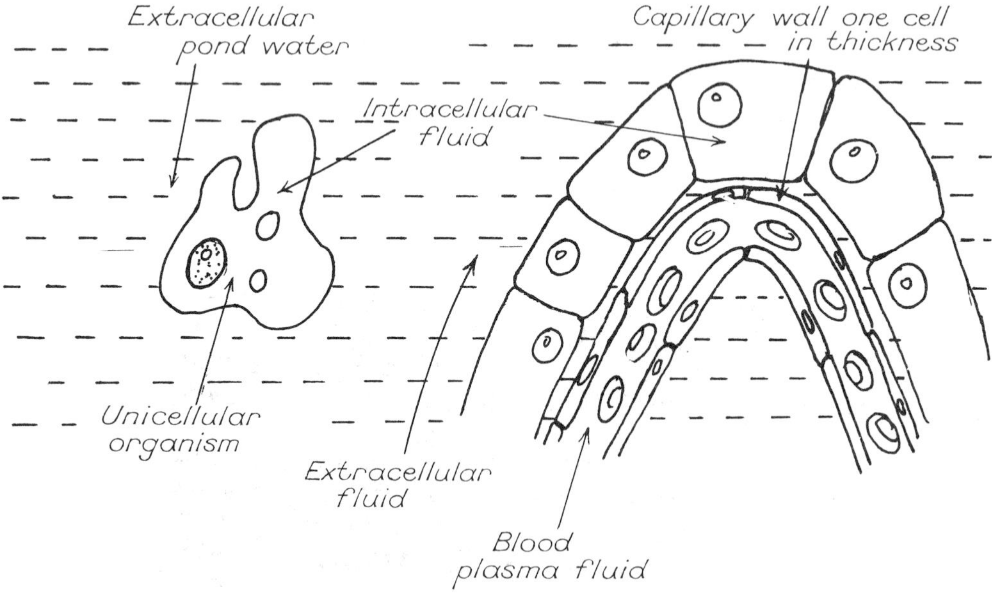

Fig. 8.9 Fluid relations for unicellular and multicellular organisms

fluid to each cell by tubular *vessels*. In plants such as ferns and flowering plants the vessels transport fluid throughout the plant; similarly in most invertebrate and vertebrate animals a system of tubular vessels called *arteries* and *veins* are in close contact with cells by way of the *capillary* vessels.

The system of circulatory vessels replenishes the supply of extracellular fluid to each cell, bringing food and oxygen, whilst removing soluble excretory waste and lowering the temperature around the cell.

TRANSPORT WITHIN CELLS

Movement of substances *within* the living cell is mainly by the process of *diffusion* in which the dissolved substance moves from a region of *high* concentration to a region of *lower* concentration. In addition the cell *endoplasmic reticulum* transports substances through the cell.

Diffusion is the method by which substances other than water enter the cell from the extracellular fluid.

Osmosis is the method of transport by which water enters and leaves the living cell through a *selectively permeable* membrane. The *cytoplasmic lining* of plant cells or the *plasmalemma* cell membrane of animal cells are such membranes. Semipermeable membranes allow *solvents*, for example water, to pass through the membrane but do not allow the large molecules of *solutes*, glucose and amino acids to pass through.

Osmotic pressure is the pressure produced when water passes through the semipermeable membrane.

Cells are in a state of *turgor* when they become rigid and cannot take in more water by osmosis, in contrast to *flaccid* cells that have lost water by either osmosis or evaporation.

The outer cellulose wall of plant cells is *permeable*: this allows the water and soluble cell substances to pass out freely, and the surrounding extracellular fluid to enter the cell.

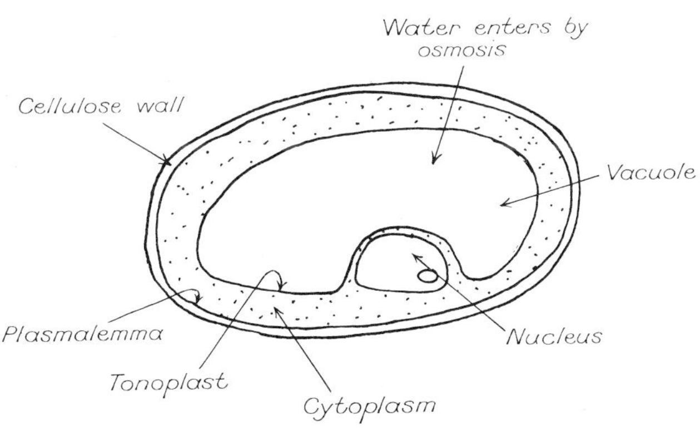

Fig. 8.10 Plant cell membranes

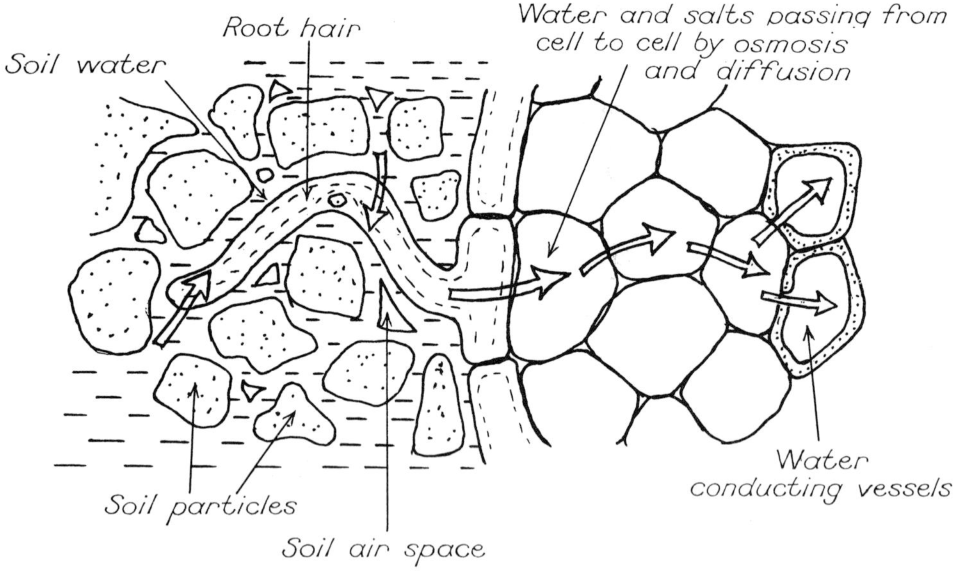

Fig. 8.11 Passage of soil water into a root by way of the root hairs

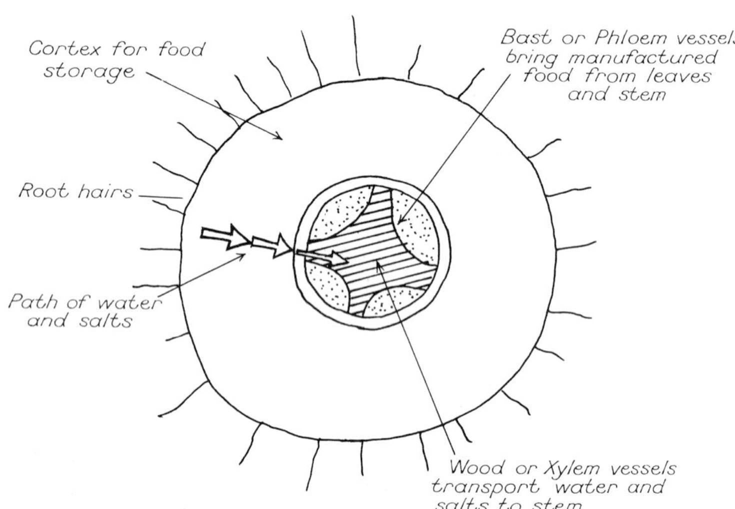

Fig. 8.12 General structure of a flowering plant root

TRANSPORT IN PLANTS

Plant *root hairs* are single cells in close contact with the soil water; millions of these root hairs provide a great *surface area* for absorption. Water enters the root hair by osmosis from the soil. Further movement of the fluids from cell to cell in the root is by osmosis and diffusion. The central cells of the root, having the *greatest* concentration of salts, cause the water to pass from cells containing the *weaker* salt solutions to the *stronger* solutions across the selectively permeable cell membrane.

Osmotic pressure is produced by the intake of water by the root hairs which causes the water and dissolved salts to pass into the conducting *xylem* vessels.

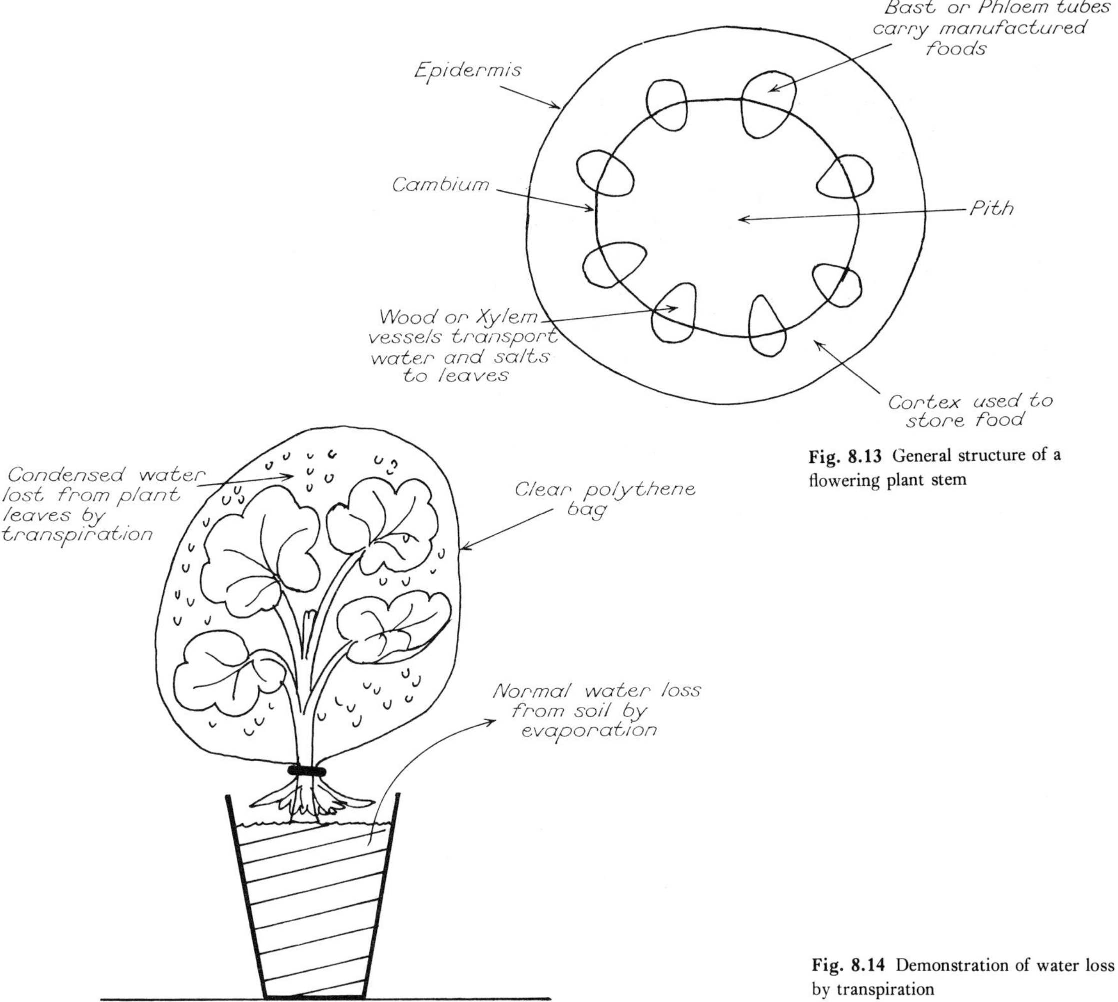

Fig. 8.13 General structure of a flowering plant stem

Fig. 8.14 Demonstration of water loss by transpiration

TRANSPIRATION IN PLANTS

Ferns, conifers, and flowering plants possess tubular *vessels* grouped in *vascular bundles* which are also called 'veins'; these serve to transport fluids throughout the plant. The water and salts absorbed by the root hairs, enter the *xylem* or wood vessels which extend from the roots up the stem to the leaves. These *wood* vessels or xylem tubes can be seen by placing a cut shoot of a flowering plant in a solution of *safranin* dye for one hour; when the shoot is cut with a sharp razor crosswise and lengthwise the distribution of the dye in the xylem vessels will be seen.

Transpiration is the movement of water through a plant from the soil via the root and stem and its loss from the leaves into the air. This is demonstrated by completely wrapping a potted plant in a polythene bag; water transpired by the leaves is seen to condense inside the bag.

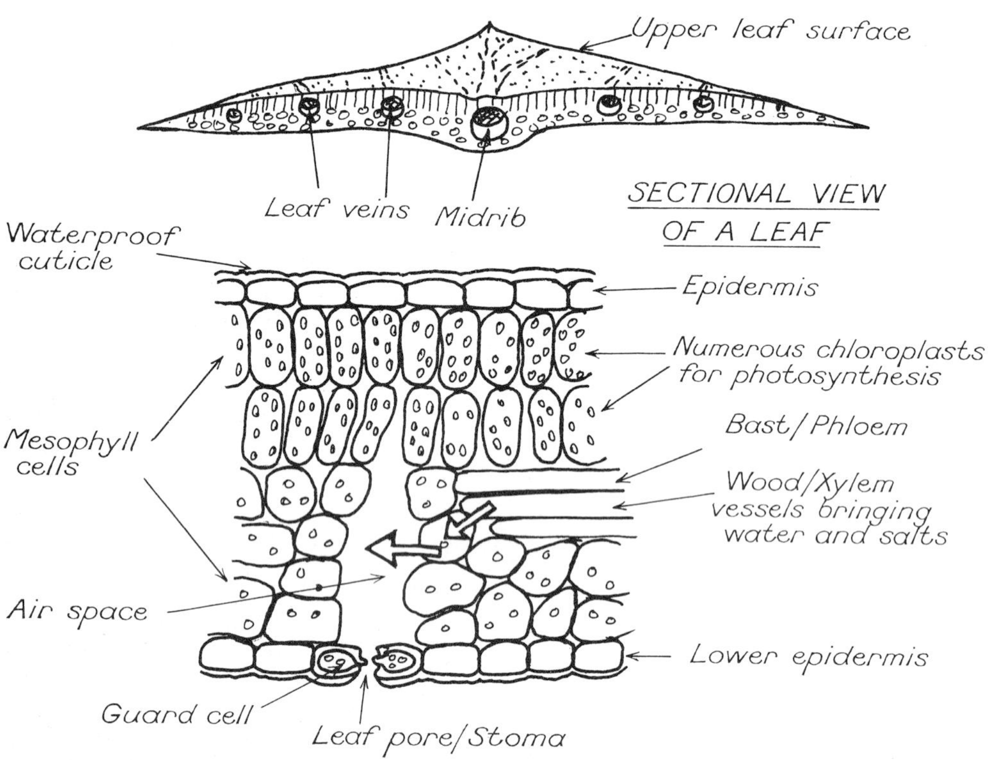

Fig. 8.15 General structure of a flowering plant leaf

Transverse sections of the leaves of a flowering plant, when examined through a microscope, are seen to have *vascular bundles* (see also Plate 29, p. 136). These bring water and salts to the *mesophyll* cells forming the interior of the leaf. Each mesophyll cell loses water by *evaporation* from its surface into the air space of the leaf, the evaporation water leaving the leaf by way of the leaf pores or *stoma* into the surrounding air. This process of evaporation causes more water to be drawn up the plant stem from the roots in the *transpiration stream.*

TRANSLOCATION

After the process of food manufacture by photosynthesis in cells which contain chloroplasts, the newly formed carbohydrate food is transported through the plant by special vessels called the *phloem* or bast. These tubular vessels complete the structure of a vascular bundle together with the xylem vessels.

Manufactured food is carried in all directions throughout the plant to the *growing regions* of the root and stem tips, or it is passed to the *storage organs* of the root, stem or seeds.

This movement of manufactured foods throughout the plant takes place mainly by diffusion, and is called translocation.

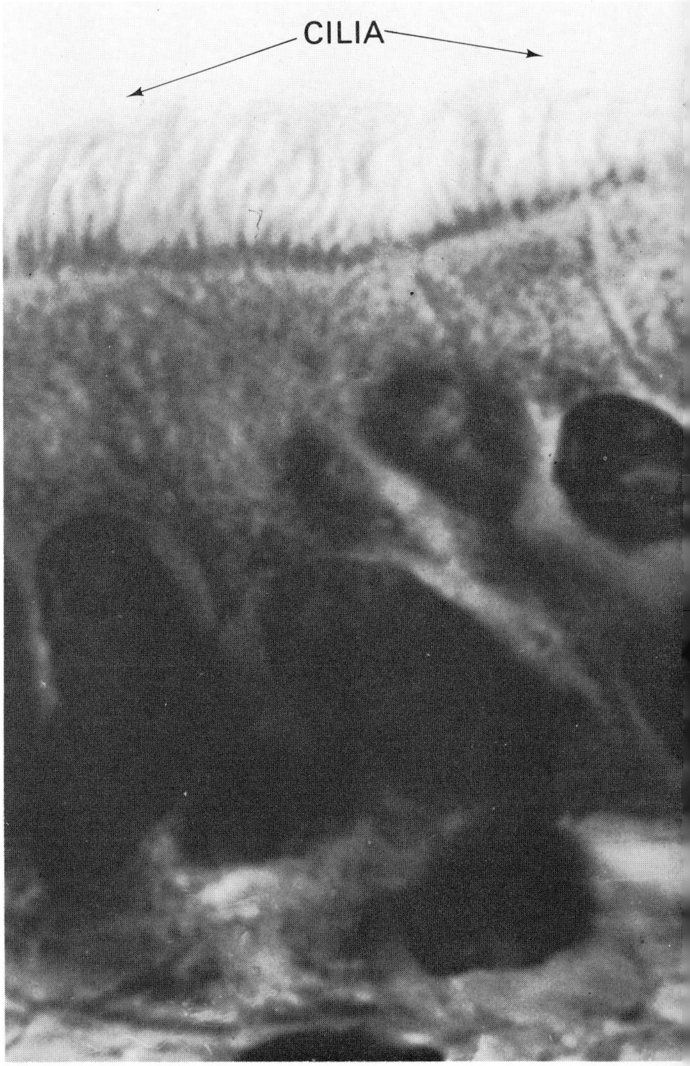

Plate 28 Epithelial cells with cilia along the top border (Griffin Biological Laboratories)

CILIA

Fig. 8.16 Ciliated epithelium

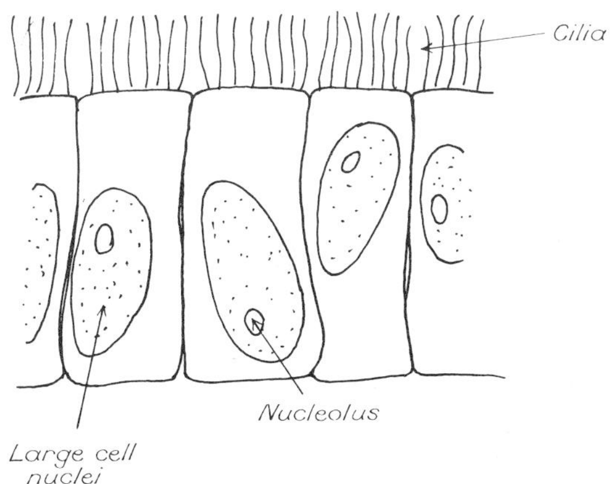

TRANSPORT IN ANIMALS

The movement of substances within multicellular animals is summarised as follows:

(a) **Ciliary movements.** Seen in cells which are equipped with *cilia*, tiny hairlike structures occurring in *epithelial* cells of earthworms, molluscs, vertebrates and man. Cilia serve to transport fluids and solid substances in the upper respiratory passages of the lung, mucus and solid dust particles being wafted forwards into the mouth. Female *oviducts* are lined with ciliated epithelium which move the ovum towards the uterus. Male spermatozoa have a whiplike *flagellum* which drives the spermatozoa towards the female ovum.

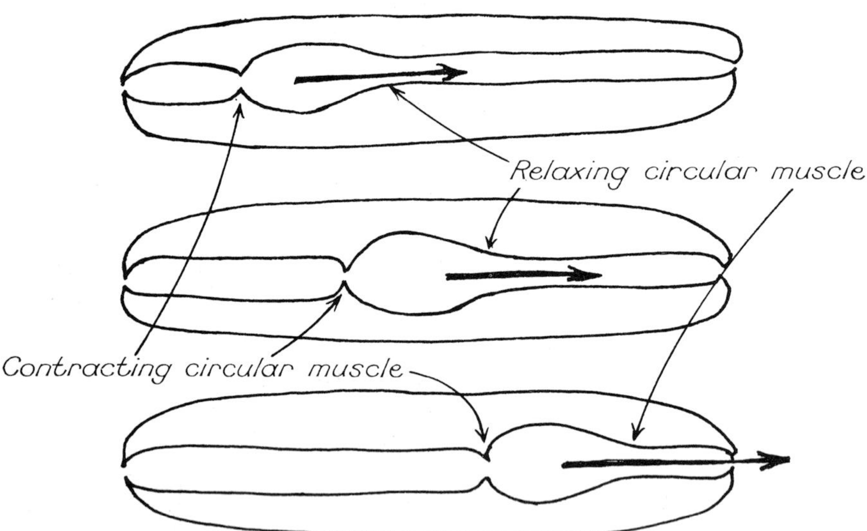

Fig. 8.17 Peristaltic movement in an animal alimentary canal

(b) **Peristalsis.** An undulating, flowing movement due to the alternate contraction and relaxation of opposing muscles called the longitudinal and circular muscles seen in the body wall of the earthworm.

Peristalsis is seen in the alimentary canal from oesophagus to the anus, and in the kidney ureters, and also in the female uterus.

(c) **Flowing movements.** These are also called *amoeboid* movements being typical of the protozoa *Amoeba*, and are seen in the white blood cells of vertebrates.

(d) **Blood.** The *plasma* fluid component of blood is brought into close contact with body cells by the capillary blood vessels. In man, body fluids consist of the *intracellular fluid* inside the body cells, and *extracellular fluids*—composed of the *interstitial* fluid around the cells and *plasma* fluid inside the blood vessel. Both the interstitial and plasma fluid are collectively called the *extracellular* fluid. In man, intracellular fluid amounts to 67%, whilst blood plasma is 6% and interstitial fluid is 27% of the total body fluids.

Invertebrates have a blood which is entirely *fluid*, whilst the vertebrates have a partly fluid and *cellular* component seen in the different blood cells.

Blood plasma, which is mostly water and contains the *solutes* amino acids, glucose, salts and urea (carbamide), provides the essential extracellular fluid needed for the exchange of substances by osmosis and diffusion.

Circulation of the blood is by means of *blood pressure* generated by the muscular heart pump; this pressure is responsible for forcing the fluid through the capillary vessel walls.

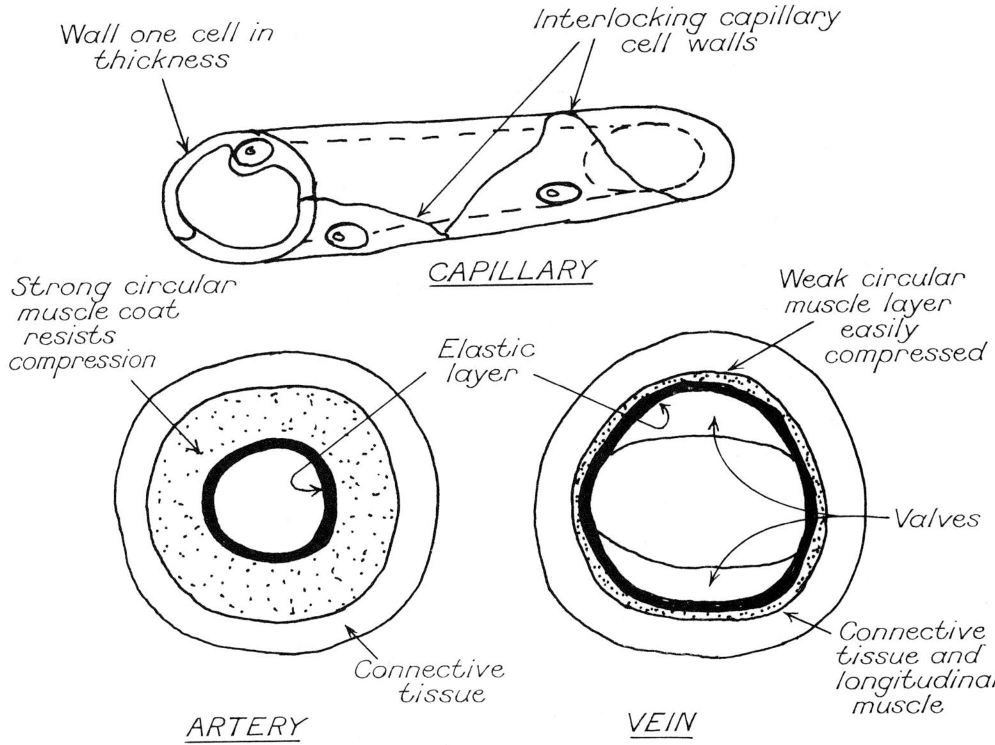

Fig. 8.18 Structure of blood vessels

Arteries, with their thick muscular walls, lead the blood away from the heart (except in the *pulmonary artery*). They link up with veins with less muscular walls, returning blood to the heart (except in the *pulmonary vein*). Back flow of blood in veins is prevented by semi-lunar non-return *valves* lining the interior of the vein.

Blood *circuits*, which are linked by blood filled spaces or *sinuses* seen in many invertebrates, are called *open* blood circuits; in contrast the *closed* circuits of the vertebrates, excepting some fish, have capillary vessels linking the artery and vein systems (see p. 49).

In addition to the heart pumping action, another force is provided by the body muscles. They compress the veins and drive the venous blood with its lower blood pressure back to the heart; this is called the *muscle pump*.

(e) **Lymphatic systems** carry the extracellular fluid through lymphatic vessels as *lymph* which is transported to the heart. The lymph moves along the vessels partly by the pulsating action of the blood vessels lying close to them. An important lymph vessel is the *lacteal* vessel of the intestinal villus, concerned with the absorption of fatty (alkanoic) acids, and glycerine (propanetriol).

FOOD STORAGE IN PLANTS AND ANIMALS

Plants transport foods in a *soluble* form to the storage organs where the food is stored in an *insoluble* form as granules of starch or protein. The formation of complex insoluble foods from the simpler soluble food materials is a chemical process called *condensation* during which a molecule of *water* is eliminated by combination of molecules to form the complex *polymers* of starch, cellulose and proteins.

Some of the main plant storage organs (shown in Fig. 12.2) include *stems* that become modified as tubers for potatoes and as *rhizomes* in many plants of medicinal value. *Roots* are the main storage organs in carrots and parsnips, whilst *seeds* store their food reserves in either an *endosperm* or within the *cotyledons* or seed leaves—for example, peas and beans.

Animals depend ultimately on plants for food, most of this plant food is used for energy, growth, tissue repair, movement and protection from certain diseases. Surplus amounts of foods are used as follows:

(a) **Carbohydrates** travel to the liver as glucose and are stored as *glycogen* or animal starch. In addition some glycogen is temporarily stored in muscles.

(b) **Lipids** form from the surplus carbohydrate in the diet, and also form glycerine (propanetriol) and fatty (alkanoic) acids in the blood. The lipids are found in fat depots beneath the skin and around the kidneys.

(c) **Protein** surplus to the body needs is changed into glycogen in the liver by a process of *deamination.*

(d) **Iron** for the formation of red blood cells is stored in the liver.

(e) **Vitamins.** A retinol, and D cholecalciferol, together with vitamin B_{12} cyanocobalamin are stored in the liver. In addition vitamin A retinol and vitamin D cholecalciferol, being lipid soluble, are found in the fat depots.

(f) **Eggs** of birds and reptiles store considerable amounts of food in the *yolk.*

(g) **Milk** produced by the mammary glands is a nutritious store of food for developing mammals.

9 Respiration in Organisms

Solar energy is stored in food as *potential energy*. This is later released for use by living organisms as *kinetic* energy. Sunlight energy is taken up by the *chlorophyll* pigment of plants and is used in the process of *photosynthesis* for making carbohydrate food material. Plants and animals release the stored potential energy to provide kinetic energy for growth, movement, nerve impulse, conduction and secretion.

Respiration is the process of releasing energy from food material in living plant and animal cells.

RELEASING ENERGY FROM FOOD

Carbohydrate, lipid, and protein foods are composed mainly of the chemical elements, *carbon* and *hydrogen*, which are able to produce energy by the chemical change called *oxidation*, or combination with oxygen.

$$\begin{array}{cccc} \text{carbon} & +\ \text{oxygen} & = \text{carbon dioxide} + \text{energy} \\ C & +\ O_2 & = CO_2 \\ \text{hydrogen} & +\ \text{oxygen} & = \text{water} + \text{energy} \\ 2\,H_2 & +\ O_2 & = 2\,H_2O \end{array}$$

The heat energy value of dry foods can be measured in laboratories in an apparatus called the *calorimeter*. The heat energy value is measured in *kilojoules* (kJ). The following are the energy values of *one gram* (1 g) portions of pure forms of the main groups of food nutrients:

One gram pure	100%	carbohydrate	17 kilojoules
		protein	17 kilojoules
		ethanol	29 kilojoules
		lipid	38 kilojoules

It is seen from the above the lipid foods provide *twice* the energy value of equal weights of either carbohydrates or protein foods, whilst ethanol gives $1\frac{1}{2}$ times the energy provided by equal amounts of carbohydrate or protein.

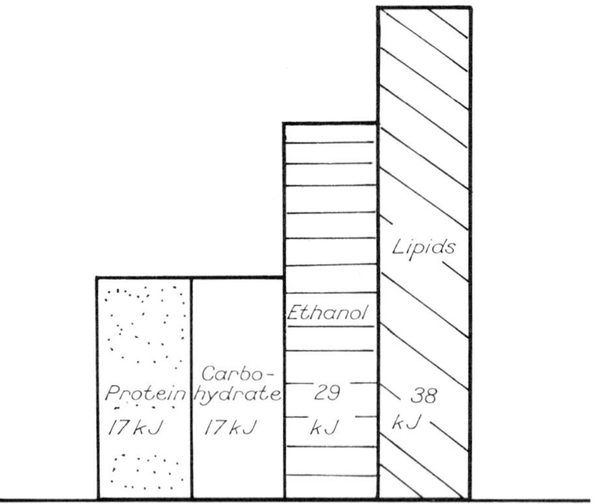

Fig. 9.1 Energy values of 1 g quantities of pure food components

ENERGY RELEASE IN ORGANISMS

The process of releasing energy from food in living plants and animals differs from the straightforward laboratory process of burning or the combustion of dry food.

The following outlines of experiments performed on living plants and animals demonstrate the process of respiration.

(1) **Carbon dioxide** is a product of respiration and its formation can be shown using the apparatus in Fig. 9.2. Air, which must be *free* from carbon dioxide, is drawn into the chamber containing either a potted plant or small animal kept in total darkness. The air, drawn out of the chamber by means of a vacuum pump, bubbles

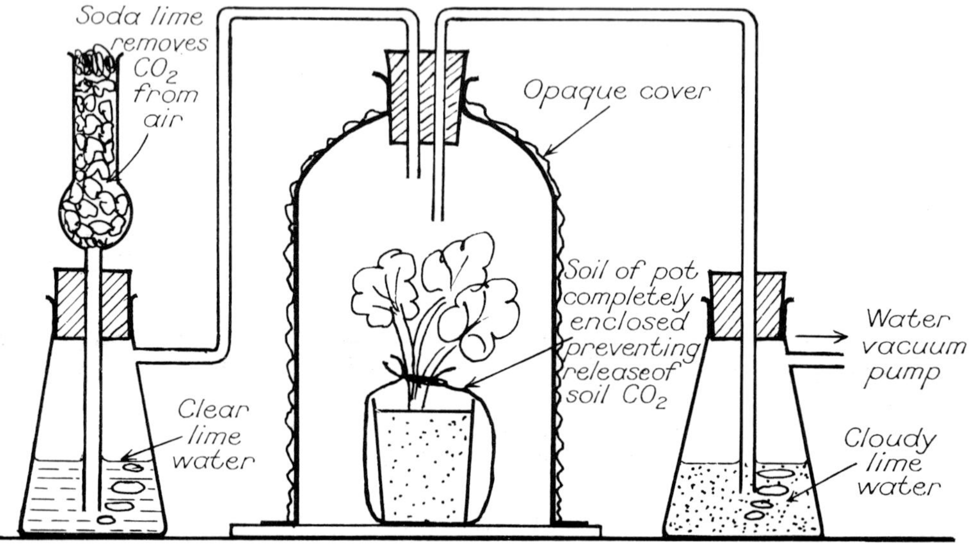

Fig. 9.2 Carbon dioxide produced by plant respiration

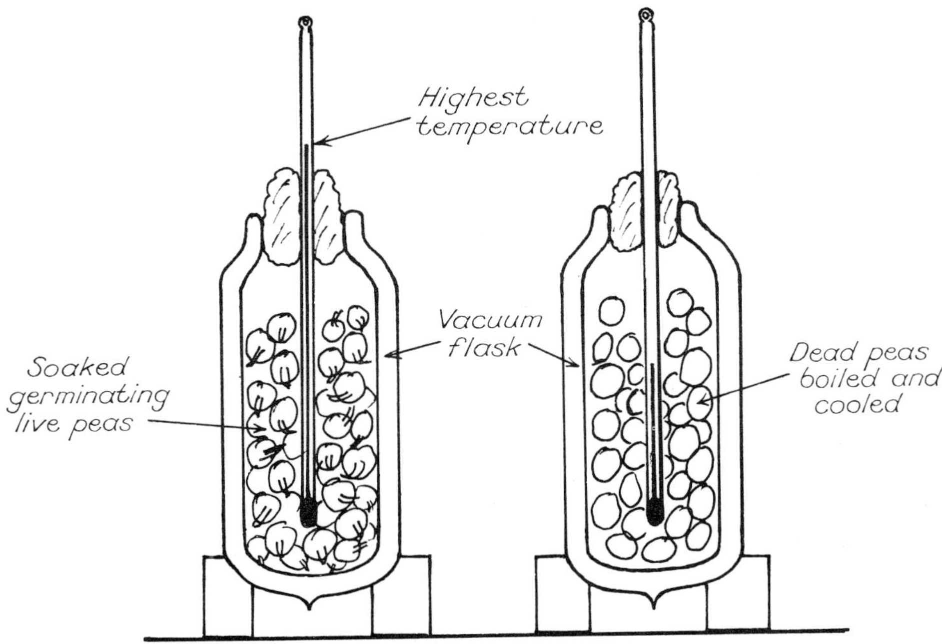

Fig. 9.3 Release of energy as heat from respiring plant seeds

through clear lime water (calcium hydroxide solution). This will turn cloudy indicating the presence of carbon dioxide as a product of respiration.

(2) **Heat** is generated within the bodies of living animals and can be detected by a thermometer. Heat generation in plants can be shown by placing soaked pea seeds to germinate in a vacuum flask. *Boiled* and cooled pea seeds, which are dead, their enzymes having been destroyed by heat in boiling, are placed in another vacuum flask. Thermometers inserted in each vacuum flask will show a rise in temperature only in the flask of germinating or live peas.

EXTERNAL AND INTERNAL RESPIRATION

Respiration is a process which will be considered under two headings.

External respiration is the process in which *oxygen* from the air is taken to the main centres of respiration, and includes the giving out of *carbon dioxide*. In plants it involves taking in air mainly by the *leaves*, whilst in animals *gills*, or tubes leading into *lungs* are used (described in earlier studies of various types of animals). External respiration is loosely described as 'breathing'.

Internal respiration concerns the release of energy from food *within* the living cells of plants and animals. This process occurs on the cell organelles called the *mitochondria*.

OUTLINES OF THE PROCESS OF INTERNAL RESPIRATION

The energy released within living cells by internal respiration comes mainly from the breakdown of *glucose* carbohydrate. The energy released can be stored in the form of an *energy-rich* chemical compound called *adenosine triphosphate* or *ATP*; this chemical compound is found in living cells close to the mitochondria.

In order to release the energy from the adenosine triphosphate (ATP), it must enter a *cycle* of chemical changes, which first involves *hydrolysis*, a chemical change involving reaction with water. This releases the energy into the cell for the various life activities. Meanwhile *adenosine diphosphate* or *ADP* has been produced; this chemical compound can pick up further energy through high energy *phosphate* bonds, reforming once again *adenosine triphosphate* (ATP), which re-enters the energy cycle.

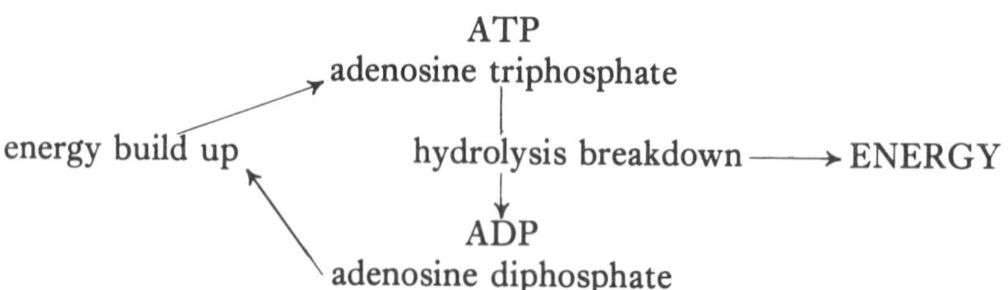

When *glucose* is broken down in the respiring cell of plants and animals, the number of very complex chemical changes which it goes through can be briefly summarised as follows:

(1) **Glucose** is changed into another chemical compound called *pyruvic acid*, in doing so energy is released and picked up by the waiting adenosine diphosphate ADP to form the energy rich compound adenosine triphosphate ATP.

(2) **Pyruvic acid** now enters a cycling process called the *Krebs cycle* during which further energy is released and picked up by adenosine diphosphate (ADP) to form adenosine triphosphate (ATP). During this cycle carbon dioxide is produced, and in addition important *hydrogen atoms* are also formed.

(3) **The hydrogen atoms** combine with *oxygen* to produce more energy and in addition water is formed

$$2\,H + O = H_2O + \text{energy}$$

The amount of energy released by this simple chemical change is far greater than the *immediate* needs of the cell. Consequently a more economical and lengthy process of energy release is performed during which the hydrogen atom is passed from one special *enzyme* to another special *enzyme*, releasing energy in *three* stages. This is gathered up by the adenosine diphosphate (ADP) to be stored as adenosine triphosphate (ATP).

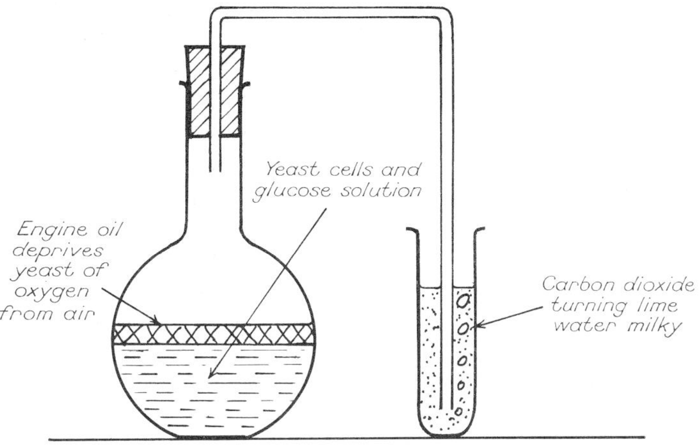

Fig. 9.4 Anaerobic respiration of yeast

(4) **Finally** the hydrogen atom is passed to oxygen by the enzyme *cytochrome* present in the mitochondria to form water, a product of the respiration process.

An important aspect of the above summary is that many stages of the respiration process are mainly controlled by *enzymes* and much energy is produced without the intervention of oxygen which only enters the process towards its end.

Anaerobic respiration

It is important to note from the previous description of internal respiration that oxygen enters the process in the very latest stages, when a good deal of the energy has been transferred to ATP by enzymes. Many animals, particularly parasites, bacteria, and fungi, can release energy from food *without the use of oxygen.* This process is called *anaerobic* respiration.

Yeast added to a solution of glucose made from cooled boiled water will have no oxygen dissolved in the water, a layer of engine oil is poured onto the solution to exclude air. The whole apparatus is then kept in a warm place for several days, a tube of lime water (calcium hydroxide) is connected to the flask as in Fig. 9.4. As carbon dioxide is released it turns the lime water cloudy, filtration of the liquid followed by distillation will produce a small amount of *ethanol* (ethyl alcohol) recognised by its smell.

The yeast cells have undergone respiration *anaerobically* without oxygen to produce energy for the cells life activity, and in addition carbon dioxide and ethanol are byproducts.

$$\text{glucose} \rightarrow \text{ethanol} + \text{carbon dioxide} = \text{ENERGY}$$

It is believed that at least eight different enzymes are involved in this process of yeast anaerobic respiration. The enzymes are collectively called *zymase.*

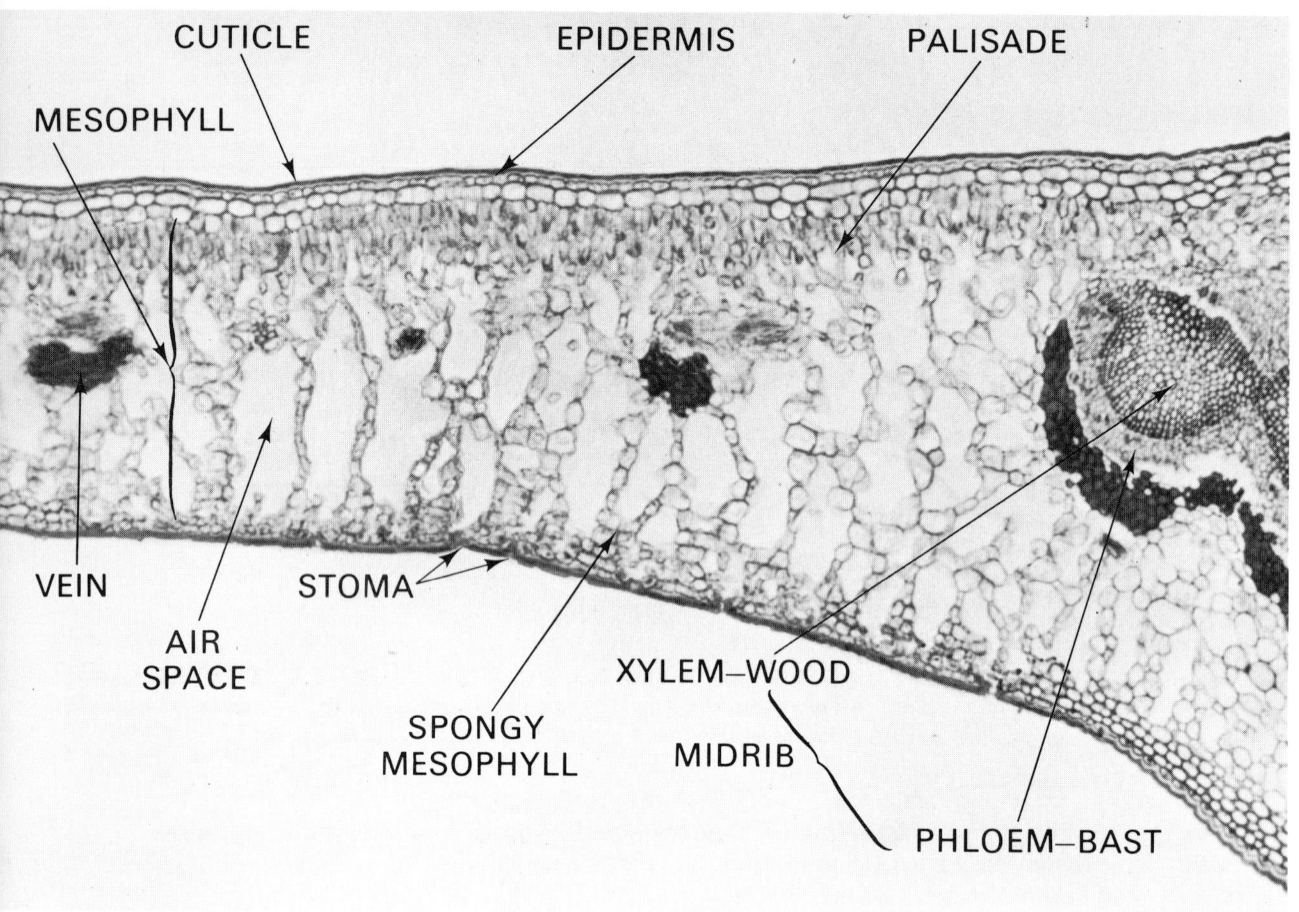

Plate 29 Sectional view of a leaf seen through the microscope (Griffin Biological Laboratories)

Anaerobic respiration occurs in the *muscle* cells of animals during intense and vigorous exercise, when oxygen is unable to reach the cells in sufficient amounts; when this occurs *lactic acid* is produced instead of ethanol as in yeast cells. On resting, the muscle cells allow oxygen to change the lactic acid into glucose which is then respired aerobically in the normal manner. If large amounts of lactic acid accumulate in the muscle, painful 'cramp' will occur. After hard muscular work, large amounts of lactic acid collect which after changing into glucose requires a high intake of oxygen for a certain period. The extra amount of oxygen required is called the '*oxygen debt*'.

Aerobic respiration

Aerobic respiration involves the use of oxygen; it is the method of internal respiration used by the majority of plants and animals, excepting the *anaerobes*, fungi and bacteria. Aerobic respiration is a much more efficient process of liberating energy from food and is considered to produce 18 times more energy than the method of anaerobic respiration.

USES OF ENERGY

Energy released during internal respiration is used for the following purposes:

(a) Movement of muscles
(b) Growth and repair of plant and animal cells
(c) Excretion and secretion of water and metabolic waste
(d) Heat for body warmth in animals and for enzyme function
(e) For the many chemical changes in metabolism
(f) Electrical energy for nerve impulse conduction
(g) Light energy seen in glowworms and certain bacteria

EXTERNAL RESPIRATION

External respiration is the process by which oxygen reaches the *mitochondria* of a living cell from the surrounding extracellular fluid or air.

CONDITIONS NEEDED FOR EXTERNAL RESPIRATION

(1) **A large surface area** is needed to allow rapid diffusion of gases into and out of each cell. Very small cells have a greater surface area to volume *ratio* and act as efficient respiratory surfaces.

(2) **Moisture** by fluids with a low surface tension allow gases to dissolve readily in order to facilitate their entry into the cell; the moisture reduces the effort needed to move the respiratory surfaces.

(3) **Thin cell walls** or membranes allow more rapid entry of gases than thicker walls or membranes.

(4) **Transport** of the gases to the respiratory organelles and surfaces can be by diffusion or air-carrying tubes and oxygen-carrying pigments.

EXTERNAL RESPIRATION IN PLANTS

Green plant cells containing chloroplasts produce oxygen as a by-product of photosynthesis; this diffuses from the chloroplast to the neighbouring mitochondria of the cell. During darkness, when photosynthesis stops, the oxygen diffuses from the air outside to reach the interior of the cell or plant. Large flowering plants have *stomata* which control the entry of air into the leaf, whilst the *cuticle* of the leaf prevents diffusion of gases in and out of the leaf (Plate 29).

Respiration takes place in the living cells of plants at all times, whilst photosynthesis occurs only during periods of exposure to light in those cells which have chloroplasts.

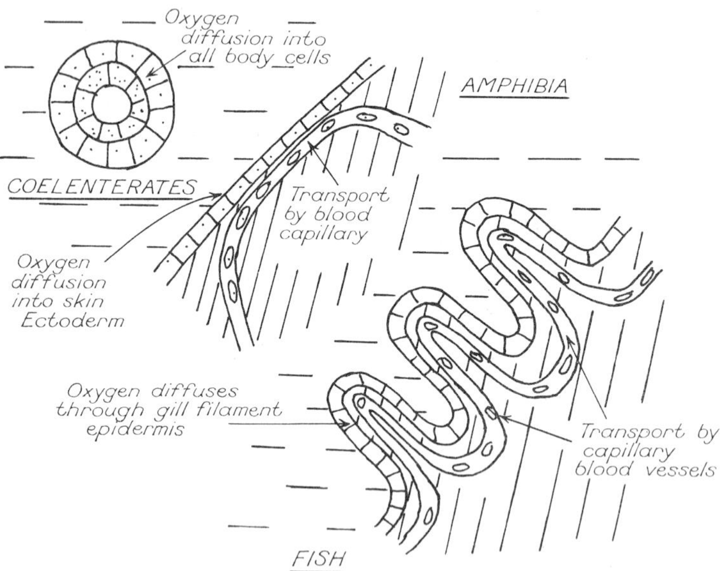

Fig. 9.5 Respiration in aquatic animals

EXTERNAL RESPIRATION IN ANIMALS

(1) **Unicellular** animals and the multicellular *coelenterates* have a large surface area to volume *ratio* and respiration is by simple diffusion of gases through the continually moist thin membranes of the cells. The biconcave shape of red blood cells of vertebrate animals are efficient for gaseous exchange because of their large surface area to volume ratio (Fig. 9.5).

(2) **Skin** offers a large surface area for respiratory purposes. This is seen in the frog and earthworm where the skin is *one cell* in thickness and is continually kept moist by a secretion of mucus from the skin cells. The skin is also well supplied with blood vessels with thin walls allowing a further rapid diffusion of gases into the blood (Fig. 9.5).

(3) **Gills** of fish, molluscs, and in the embryo stages of amphibia, consists of *filaments* kept continually moist by immersion in water, allowing diffusion of gases into the blood capillary vessels through the thin walls (Fig. 9.5).

(4) **Air tubes** or tracheae, seen in insects, carry the air *direct* to the cells bathed in an extracellular fluid (Fig. 9.6).

(5) **Lungs** are thin walled sacs which are inflated and deflated by respiratory movements of the body wall. Frequent movement of the respiratory surface is involved and a secretion of the lung lining produces a *detergent* fluid with a low surface tension which helps to reduce the effort needed in the inflation and deflation of lungs (Fig. 9.6).

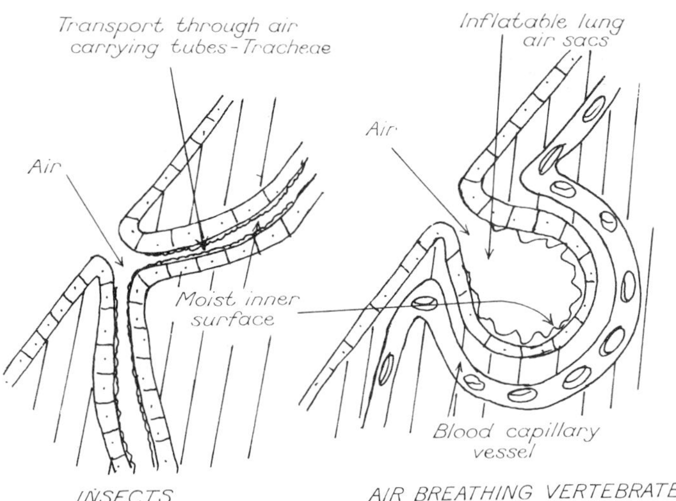

Fig. 9.6 Respiration in air breathing animals

GAS EXCHANGE BETWEEN THE BLOOD AND BODY CELLS

After oxygen has diffused into the blood capillary system it commences its transportation to the body cells as follows:

(a) **Haemoglobin** in the red blood cells readily forms an *unstable* chemical compound called *oxyhaemoglobin*; by this means the oxygen content of capillary blood becomes 20%, in comparison to the 21% oxygen content of the air in the lung air sacs. Owing to the higher partial pressure, oxygen will diffuse into the blood.

(b) **Oxyhaemoglobin** is bright red in colour, compared to the darker red colour of haemoglobin in the veins.

A molecule of haemoglobin (Hb) can combine with *four* molecules of oxygen as follows:

$$\text{haemoglobin} + \text{oxygen} = \text{oxyhaemoglobin}$$
$$Hb + 4\,O_2 = HbO_8$$

The unstable oxyhaemoglobin breaks down releasing its oxygen and reforming haemoglobin to act further as an oxygen 'carrier'.

$$\text{oxyhaemoglobin} = \text{haemoglobin} + \text{oxygen}$$
$$HbO_8 = Hb + 4\,O_2$$

In contrast the poisonous gas *carbon monoxide* forms a *stable* compound called *carbonyl haemoglobin.*

(c) **Tissue fluid** is essential to dissolve the oxygen, allowing it to pass readily by diffusion through the cell to reach the cell organelles of the mitochondria.

EXCRETION AND ELIMINATION

Catabolism is the process of breaking down food to produce energy and the various chemical *waste* products. The removal of these mainly toxic chemical waste products is called *excretion.*

Defaecation is a process of removing indigestible food materials, forming the faeces, which have not entered the metabolic process; it is *not* a process of excretion. Defaecation is seen in the ejection of faecal waste from the food vacuoles of protozoa, and the archenteron of the coelenterates. Animals with a tubular gut eject the faecal waste through the anus.

A few plants which feed on insects, the *insectivorous* sundew and butterwort, trap insects on their leaves and, after digesting their food nutrients, leave the faecal waste on the leaf surface.

The main excretory products of living organisms are *carbon dioxide, water,* and certain compounds of *nitrogen.*

excretory products

Plants: carbon dioxide, water, compounds of nitrogen, alkaloids, oils and resins, and acids

Animals: carbon dioxide, water, compounds of nitrogen—as urea (carbamide), uric acid, chloride salts, phosphates and coloured bile pigments

WATER LOSS IN LIVING ORGANISMS

Water is a continually produced byproduct of metabolism in plants and animals. It is also an important circulatory *vehicle* in the transpiration process of flowering plants, and circulation of blood in animals.

Water constantly enters and leaves the plant and animal body. The following summary shows how this loss of water can occur:

(1) **Temperature.** A rise in temperature around the plant or animal will cause an increase in the transpiration or the sweating process, both of which will result in a cooling of the plant or animal.

(2) **Wind** will remove moisture from around the exposed surfaces of the living organism; this is prevented by a waterproof cuticle in flowering plants, or greasy sebum in mammal skin.

(3) **Humidity,** or the amount of water vapour in the air, affects water loss. High humidity reduces transpiration in plants, whereas mammals secrete sweat which fails to evaporate and cannot therefore produce the cooling effect. The dehydration of internal respiratory epithelia will occur if oxygen and anaesthetic gases are not suitably humidified.

(4) **Kidneys** excrete most of the water lost from vertebrate animals, being lost mainly via the lungs and to a lesser extent in the faeces.

Dehydration of the body can occur by excessive vomiting, diarrhoea and kidney action.

EXCRETION IN PLANTS

The loss of carbon dioxide and water through the *stoma* of leaves is the main excretory method of plants. Many plant excretory materials are also chemical substances of medicinal value, for example the *alkaloids*. Plant excretory products collect in the plant in a form which renders them harmless to the plant cells, as oils, resins, and as the distasteful nitrogen compounds found in the bark of some plants which help to protect it from being eaten by grazing animals.

EXCRETION IN ANIMALS

Many animals have special excretory organs to remove the soluble excretory product. The following is a summary of them:

(a) **Protozoa and coelenterates** excrete by simple diffusion into the *excretory vacuoles* and outward diffusion to the exterior.

(b) **Earthworms** have special *nephridia* or ciliated funnels collecting waste from the *coelom* space and passing it through tubes supplied with blood vessels able to remove soluble waste.

(c) **Insects** have *malpighian tubules* suspended in blood spaces of the coelom, the tubules removing waste from the blood and passing it into the alimentary canal. It is eliminated mainly as crystals or urea (carbamide) and uric acid.

(d) **Vertebrates** eliminate waste by several methods:

(1) *Kidneys* which are efficient blood filters, closely linked with the coelom, and function by diffusion, osmosis, and selective absorption.

(2) *Skin* sweat glands can produce salts, urea (carbamide) and water as excretory products.

(3) *Lungs* eliminate water and carbon dioxide.

(4) *Colon* of the large intestine absorbs water and certain salts from the faeces.

(5) *Liver* eliminates waste bile pigments by way of the bile duct, gall bladder and gut. Urea (carbamide) is formed from the *deamination* of protein.

10 *Life and the Environment*

Living organisms have *two* environments, the *internal* environment inside the cells and the *external* environment around the cells.

Unicellular plants and animals have their internal environment separated from the external environment by the cell wall or membrane, whilst multicellular organisms have the deep inner cells separated by a skin or protective epidermis.

Close contact is made between the two environments through the *extracellular fluid* and by means of a sensory *nervous system.*

THE EXTERNAL ENVIRONMENT

The earth provides a *physical* environment which affects living things through *water* or humidity, *salinity* or salt content of seawater, *temperature*, *light* and *type of soil.*

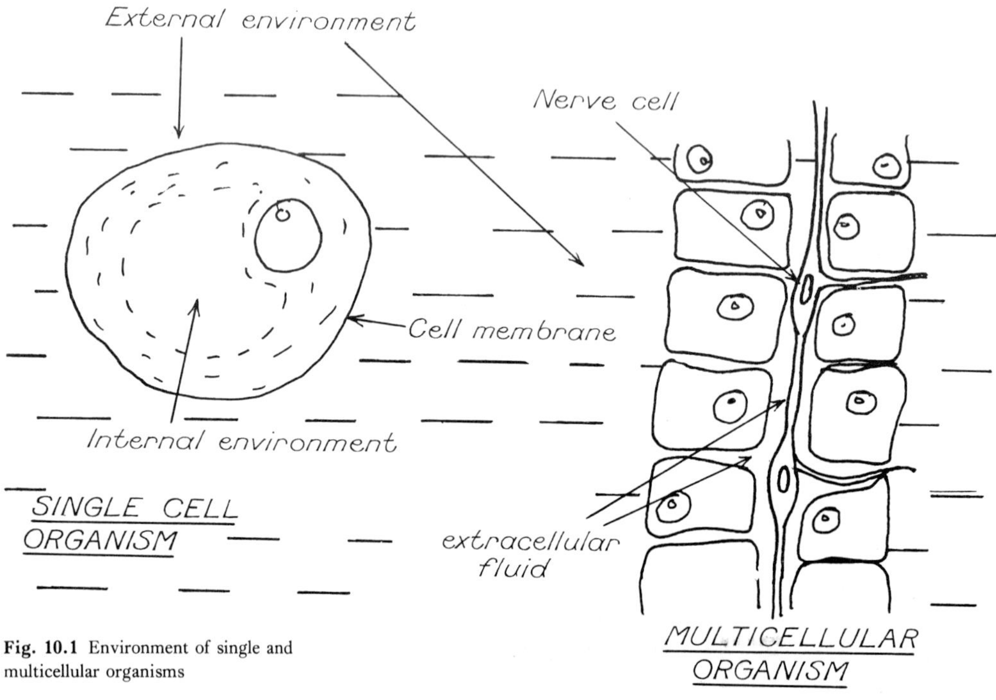

Fig. 10.1 Environment of single and multicellular organisms

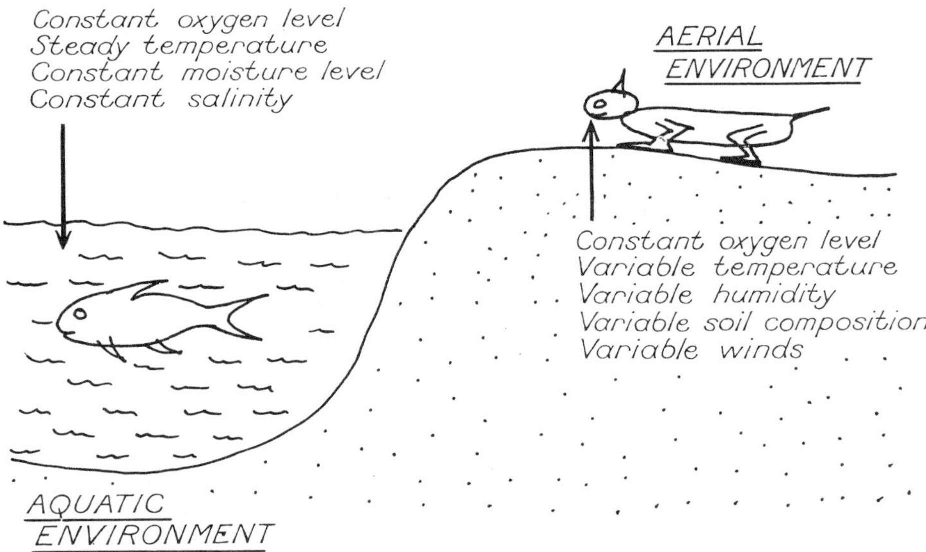

Fig. 10.2 Comparison of aquatic and terrestial environments

(a) **Freshwater and seawater environments** of *aquatic* plants and animals show a fairly *stable* or steady condition, with little change in water content, oxygen supply, salinity, pH, and pressure. These are conditions that do not change rapidly; if changes do occur they take place slowly, giving the organism an opportunity to adapt.

This stable aquatic environment is seen in the embryonic development of man, where the fetus is within the aquatic environment of the amniotic fluid.

(b) **Land or aerial environment** has only one stable environmental condition in the constant level of *oxygen*, whilst most other environmental conditions can change rapidly. This provides an *unstable* physical environment to which the organisms must adapt itself.

Birds and mammals have internal controls which make them *homoiothermal* to counteract the changing temperature of their environment, whilst most plants show dormancy in seasonal changes.

THE INTERNAL ENVIRONMENT: HOMEOSTASIS

In order that the metabolic processes which make up the internal environment activity proceed at a steady rate, selfregulating controls called *homeostats* operate and result in *homeostasis* or a steady internal environment.

The varying activities of millions of cells composing the multicellular organism are coordinated and controlled through homeostasis in *water balance, body temperature, respiration,* and *blood sugar* level. In addition changes in muscle tone and the position of the animal body in space can be detected.

ENZYMES

Enzymes are the biological catalysts which regulate and bring about *specific* chemical changes inside and outside the cell during respiration and digestion.

The activity of enzymes is closely affected by changes in the internal and external environment as follows:

(a) **Temperature.** In mammals enzymes operate best at blood heat 38°C; if the temperature falls below this value as in *hypothermia* the enzyme activity slows down. Temperatures above 50°C will increase enzyme activity until higher temperatures are reached when they are destroyed.

(b) **Poisonous chemicals** such as arsenic, lead and strychnine attack the enzymes and make them unable to function. Waste products of metabolism if not efficiently removed can also affect the enzymes.

(c) **pH.** Most enzymes function best at about pH 7 to 8.

(d) **Water.** Since many enzymes are involved in *hydrolytic* changes and since water is the essential solvent for chemical changes to take place, a disturbance in the water balance will affect enzyme activity.

(e) **Coenzymes.** These substances, such as the vitamins of the B group, as well as the metals copper, magnesium, and molybdenum, are needed for enzymes to function correctly. Thus the need for attention to these mineral substances in a balanced diet.

There are many types of enzymes: for example the enzyme *catalase*, found in living cells, is concerned with releasing oxygen from hydrogen peroxide, which forms during metabolic activity. If pieces of *fresh* carrot or meat are added to a solution of 5% hydrogen peroxide, bubbles of oxygen gas will be evolved. A similar reaction occurs when hydrogen peroxide antiseptic is applied to a bleeding wound. Pieces of boiled carrot or meat do not release oxygen from hydrogen peroxide since the *catalase* enzyme has been destroyed by the heat.

ENZYME TYPES

(i) *Hydrolases* are enzymes catalysing *hydrolysis* the change mainly occurring during the digestion of food.

(ii) *Transferases* are concerned with enzyme action in excretion, and deamination of proteins.

(iii) *Oxidases* catalyse processes of oxidation in the mitochondria.

(iv) *Fermentation* enzymes remove carbon dioxide in the anaerobic respiration process.

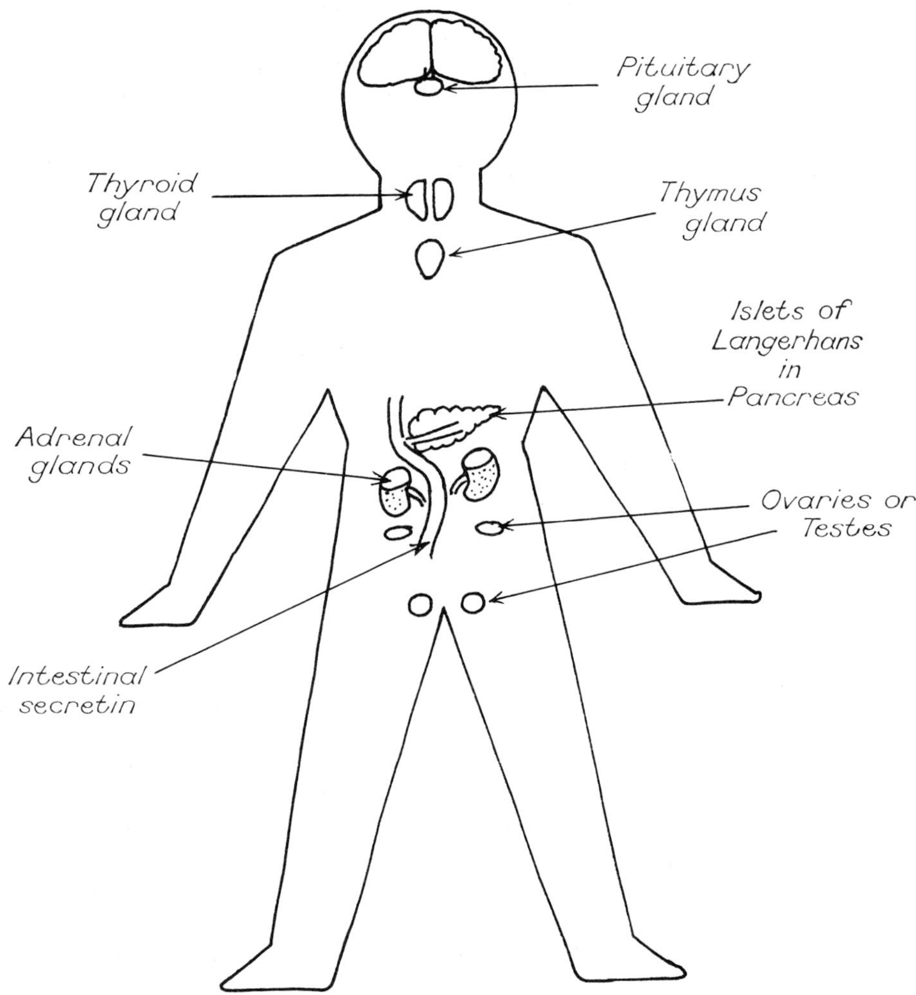

Fig. 10.3 Hormone ductless glands in man

HOMEOSTATIC CONTROLS IN MAN AND ANIMALS

Coordination of cell activities in animals is through the *nervous system* and by *chemical* control through *hormones* which are chemical messengers travelling through the blood, but are not as rapid as the messages which pass along the nerves.

HORMONES AS CHEMICAL CONTROLS

Vertebrates have two types of *glands*: *duct* glands secreting into a tubular duct (as in salivary glands) and *ductless* or endocrine glands secreting directly into the *blood system* (such glands include the thyroid and adrenal glands). The pancreas and the reproductive organs are able to produce hormones although they are classified as glands with ducts.

Hormones can control more than one process, in contrast to enzymes which are specific in their action. The following summarises the major animal hormones and their functions.

Hormone source	Location	Function
Pituitary	At the base of the brain	(1) Master gland controlling other ductless glands (2) Kidney and water balance control (3) Growth and calcium control (4) Milk secretion (5) Involuntary muscle control
Thyroid	At the base of the neck	(1) Controls development (2) Respiration control (3) Energy release from food (4) Mental activity
Adrenal	Above the kidneys	(1) Blood sugar, blood pressure, and pulse control (2) Controlling salinity of urine
Parathyroid	Partly embedded in the thyroid gland	Controls calcium levels in body
Pancreas	In parts called the 'islets of Langerhans'	Insulin produced to maintain balance of glucose in blood and formation of glycogen
Intestine	Duodenum lining	Produces the hormone secretin which activates the pancreas to release its digestive juices

HORMONES IN PLANTS

Flowering plants produce hormones in the cells found at the *tips* of the root and stem. These hormones called *auxins* are concerned with plant movements; this is seen in the bending of stems towards light and of the movement of roots towards water supplies. The flowering times are also affected by plant hormones.

HOMEOSTASIS IN FLUID BALANCE

Aquatic animals are unable to survive out of water since they have only limited means of controlling water loss. Land animals control the intake and output of water by homeostasis. Water is lost by the skin, lungs, kidneys and can also be lost by *vomiting*, *diarrhoea*, or *aspiration* of fluids; this water loss causes an *increase* in the salt concentration in the blood which is detected by *salt sensitive* homeostats in the brain.

(a) The salt sensitive homeostats cause the pituitary gland to release the *antidiuretic* hormone, ADH, which causes the kidney to reabsorb water.

(b) Further sensory nerve cells in the brain set off a mechanism causing *thirst* which, if quenched, results in dilution of the blood and restores the salt concentration to normal.

(c) Changes in the *volume* of the blood fluid are detected by sensory receptor nerves, called *baroreceptors*, located in the heart and blood vessels, which either prevent or encourage water loss. An increase in blood volume for example will cause stoppage of the antidiuretic hormone ADH from the pituitary gland.

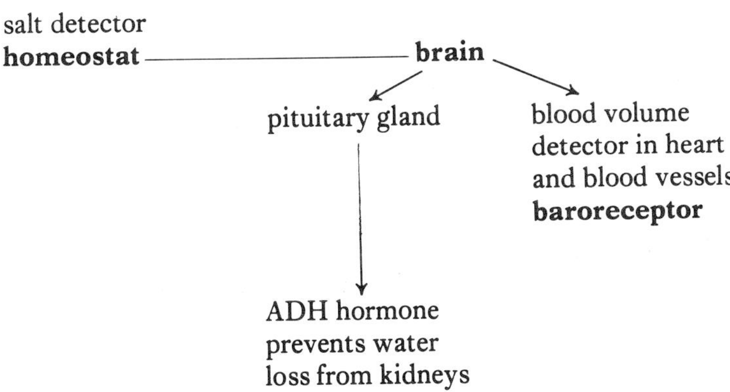

In water homeostasis there are two homeostats operating, the salt detectors and the baroreceptors.

HOMEOSTASIS IN BREATHING CONTROL

Cells in the brain are sensitive to changes in the concentration of *carbon dioxide* of the blood, which also causes a change in *blood pH*. Increasing concentrations of carbon dioxide in the blood cause the rate of breathing to increase. Other detectors sensitive to the lowering of the *oxygen* content are found attached to certain arteries; in addition the lungs possess stretch receptors or *proprioreceptors* which link with the brain respiratory centre.

HOMEOSTASIS IN BODY TEMPERATURE CONTROL

Cold blooded *poikilothermic* animals are active in a hot external environment and sluggish in a cold environment. Warm blooded *homoiothermic* animals have self regulating homeostats or *thermostats* which maintain a constant body temperature (at 38°C in man). The 'thermostat' or heat sensitive centre is located in the brain in the region of the *hypothalamus*. It can be affected by various drugs which stop the shivering action of the body in its attempt to *raise* the body temperature; such drugs are given in surgical *hypothermia*. Aspirin is also able to affect the hypothalamus to reduce body temperature in feverish illness.

The brain 'thermostat' also controls the blood capillaries in the skin causing them to dilate in *vasodilation* by relaxation of the blood vessel muscular wall, and to constrict in *vasoconstriction* by contraction of blood vessel muscular wall.

Sweat glands are controlled by nerves from the hypothalamus of the brain; the evaporation of sweat causes the cooling of skin.

Shivering, which is the rapid movement of muscles with consequent heat production, is also controlled by hormones of the thyroid and adrenal glands.

HOMEOSTASIS IN BLOOD SUGAR CONTROL

Living cells of animals need glucose to liberate energy by respiration; nerve cells also require a supply of glucose. If the heart stops beating for more than two minutes, damage to the brain and nerve cells may arise due to glucose and oxygen deprivation.

(a) **Carbohydrate foods** produce glucose on digestion.

(b) **Living cells** steadily use up the supply of blood glucose, and a good supply must be continually available.

(c) **Control** of the glucose level in the blood is by the hormones *insulin* and *adrenalin.*

A high concentration of blood sugar causes the pancreas to release insulin which changes the glucose into *glycogen.*

A low concentration of blood sugar, causes the *adrenalin* to change glycogen into *glucose.*

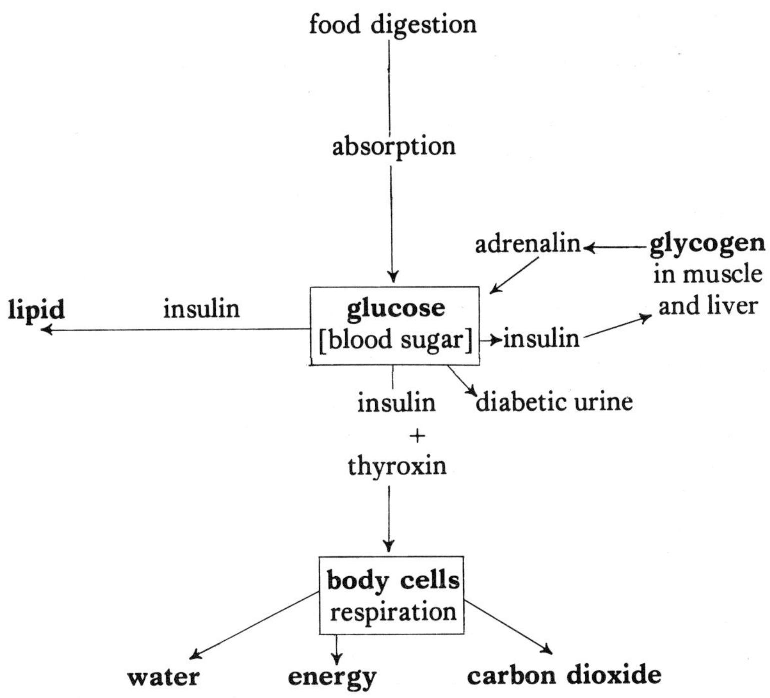

(d) **Glucose** is rapidly used up by living cells in the presence of insulin and thyroxin.

(e) **Proteins** can also be changed into glucose by other adrenal gland hormones.

NERVOUS CONTROL

Changes in the external environment are made known to the living animal through its *nervous system.*

The main environmental stimuli affecting living organisms are:

(a) **Chemicals** detected by smell and taste in animals.

(b) **Electrical stimulus** affecting muscle.

(c) **Heat, light and gravity** are stimuli affecting both plants and animals.

(d) **Pressure and sound** affecting most animals.

Living organisms *respond* to the external stimuli through *movement* or by *glandular secretion*; the reaction of the organism to the stimulus is known as *behaviour.*

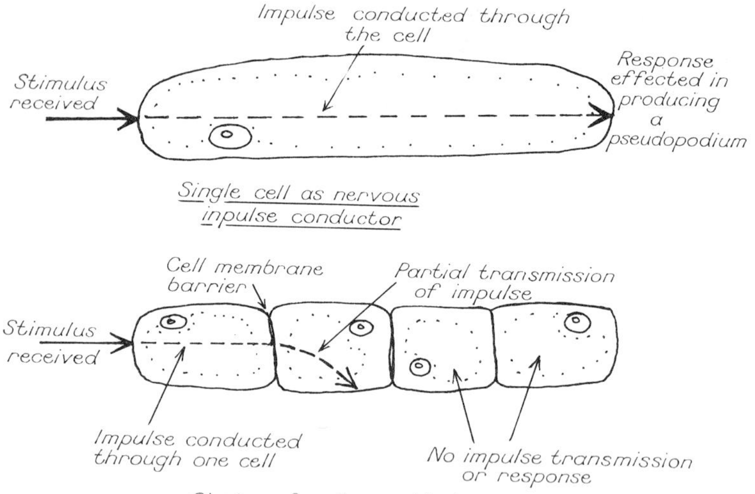

Fig. 10.4 Conduction of nerve impulses in cells

NERVE CELL OR NEURON

A unicellular animal can receive a stimulus at one point of its ectoplasm and *conduct* it through the *cytoplasm* as an *impulse*; when it reaches the opposite end of the cell a *response* is produced as either the formation of a pseudopodium or the movement of flagella or cilia, causing the animal to *move* away from or towards the stimulus (Fig. 10.4).

A multicellular animal is unable to conduct the impulse through several neighbouring cells as the cell membranes act as barriers and resist the flow of the conducted impulse. Consequently the stimulus would not reach the distant cells and little if any response would be effected. A special system of nerve cells or *neurones* is developed in

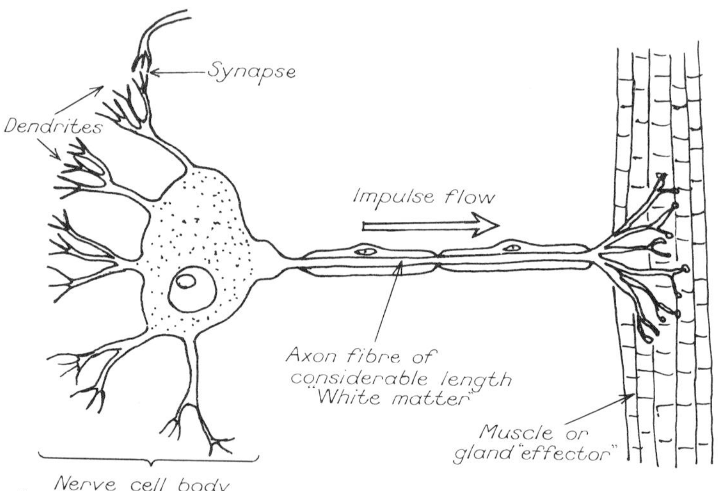

Fig. 10.5 A nerve cell or neuron (also spelt neurone)

multicellular animals; this serves to conduct stimuli and co-ordinate the activities of several cells.

A neuron (also spelt *neurone*) is composed of a nerve *cell body* with a nucleus, and the nerve *axon fibre* which can be up to a metre in length and terminates in a branched *dendrite* linking up with other neuron dendrites through a *synapse*. The stimulus is conducted *electrically* along the axon fibre (Fig. 10.5).

ANIMAL NERVE SYSTEMS

Neurones in multicellular animals are arranged either as *nerve nets* or as *central nerve systems*.

(a) **Nerve nets** are seen in coelenterates and conduct the stimuli in a general way all over the body producing a general reaction to the stimulus. A nerve *plexus* is a network of nerves found in vertebrate animals.

(b) **Central nervous systems** consisting of a brain and nerve cord are located in a *ventral* position in the *invertebrates*, whilst the brain and nerve cord of vertebrates are located in the *dorsal* position. In both the brain develops in the *anterior* position as this part of the body moves first and receives all initial stimuli.

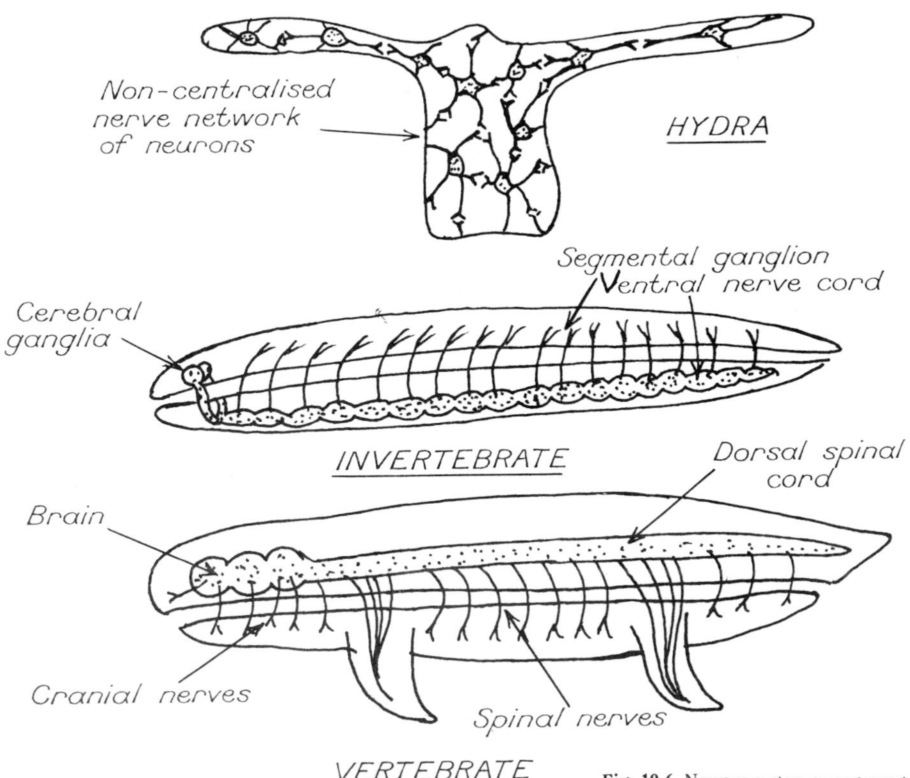

Fig. 10.6 Nervous system arrangement in different animals

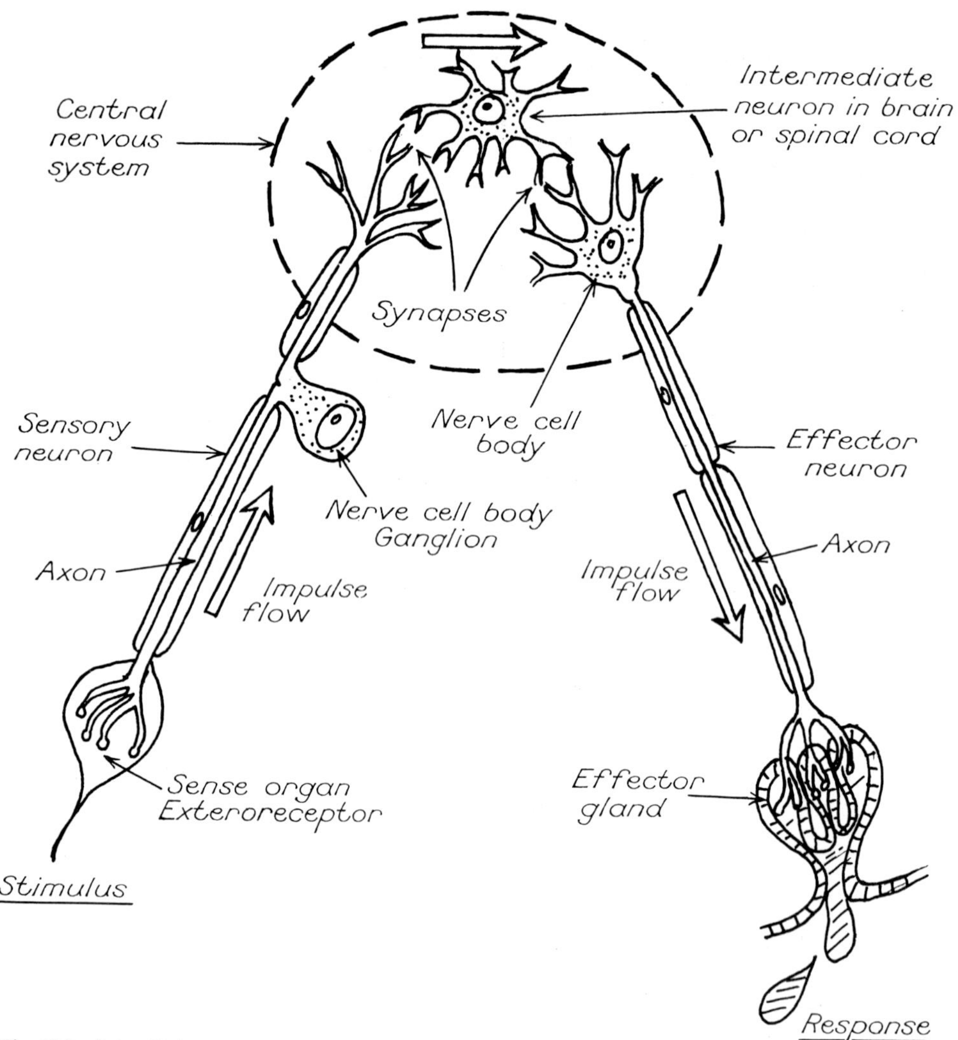

Fig. 10.7 relationship between sensory and motor nerves

Structure of the nervous system

A vertebrate's nervous system has the following basic structure:

(a) **Receptors** which receive the external stimulus that is conducted as an impulse through a *sensory* nerve to the central nervous system.

(b) **A central nervous system** of the brain or the nerve cord which may alter or interpret the impulses from receptors.

(c) **Effectors** which are muscles or glands, receiving the impulses along *motor* nerves.

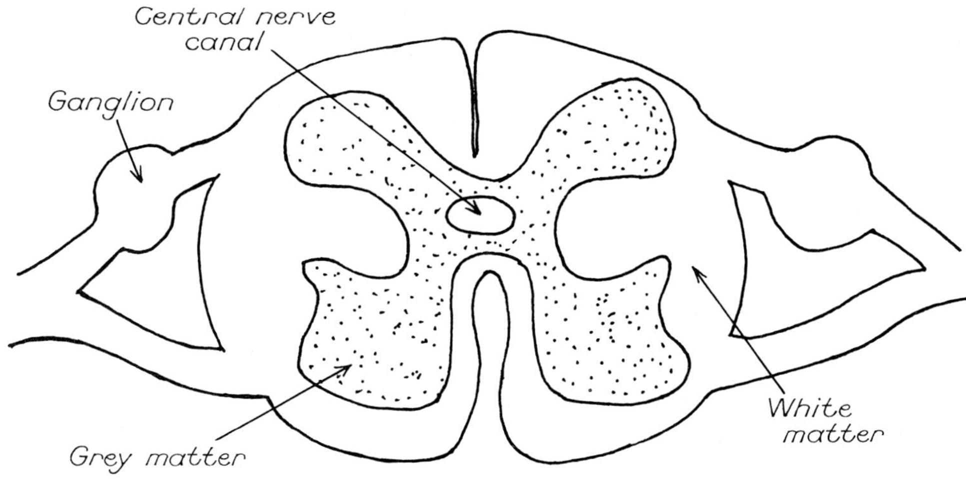

Fig. 10.8 Spinal cord structure

Receptors are of two main kinds: *external* receptors or *exteroreceptors* in the sense organs of the skin, eyes, nose, tongue and ears; and *internal* receptors or *proprioreceptors* in muscles, joints, ligaments and also linked with blood vessels.

CENTRAL NERVOUS SYSTEM

Grey matter of the central nervous system consists of the neurone nerve cell bodies, whilst the *white matter* is composed of the axon nerve fibres.

The brain grey matter is found on its *surface*, whilst the spinal cord has its grey matter in a central position as shown in Fig. 10.8. The numerous infoldings or convolutions of the brain allows a greater number of nerve cell bodies to be accommodated which is essential in animals, mainly mammals, which display intellect and memory.

Effectors from the central nervous system control the striped voluntary muscle of the arms and legs, whilst an *autonomic* nervous system connects with certain glands and involuntary unstriped muscle of blood vessels, bladder, gut wall and the eye:

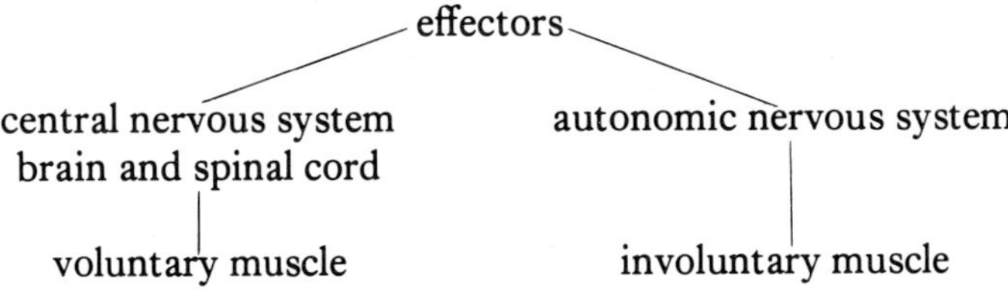

MUSCLES AND NERVES

When a nerve impulse travels along the axon nerve fibre there is a rapid exchange of *sodium* ions in the extracellular fluid for

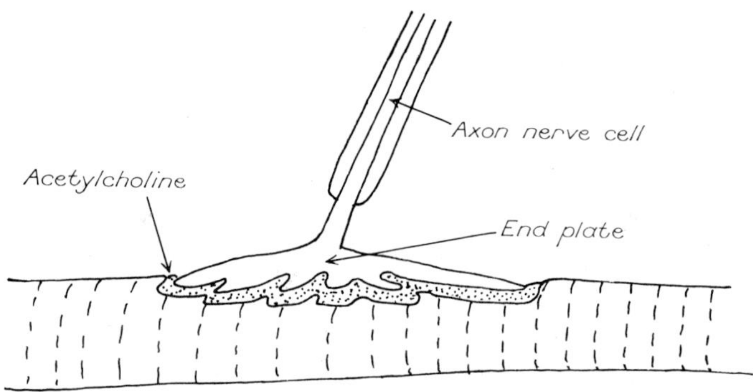

Fig. 10.9 Nerve endplates

potassium ions in the intracellular fluid. The electrical current, on reaching the dendrite ending acts on a *nerve end plate* attached to the muscle fibre. The electrical stimulus releases a chemical substance called *acetyl choline* which transmits the impulse across the synapse. The muscle *contracts* due to energy provided by *adenosine triphosphate* ATP.

Soon afterwards the acetyl choline is rapidly changed by an enzyme into *choline* and *acetic (ethanoic)* acid which allows the muscle to *relax*.

Muscle relaxants include such drugs as *curare* and *atropine*, and act by blocking the action of acetyl choline. This allows the muscle to be in a relaxed condition for surgery. Opposite acting drugs such as *physostigmine* allow acetyl choline to remain unaffected by the enzyme; this drug is used to contract the iris eye muscle in order to dilate the pupil.

AUTONOMIC NERVOUS SYSTEM

Autonomic or *automatic* nervous systems control and co-ordinate the involuntary muscles. *Two* effector or motor nerves supply the involuntary muscle fibre. One effector nerve is called the *sympathetic* nerve; in most organs, it causes stimulation of the muscle by releasing a chemical substance called *noradrenalin*—such nerves are adrenergic. The other effector nerve is called the *parasympathetic* nerve and is inhibitory in its action; it produces acetyl choline which affects the muscle in the opposite manner to the sympathetic nerve—nerves which produce acetyl choline are cholinergic.

Lungs have muscles in the bronchi which are relaxed by *noradrenalin* from sympathetic nerves; this allows the bronchial tubes to dilate. Secretion of *acetyl choline* by parasympathetic nerves causes the muscle to contract and the bronchial tubes narrow.

Most glands have *one* of the autonomic nerves. Sweat glands have only a sympathetic nerve, whilst the gastric, pancreatic, and salivary glands have a parasympathetic nerve.

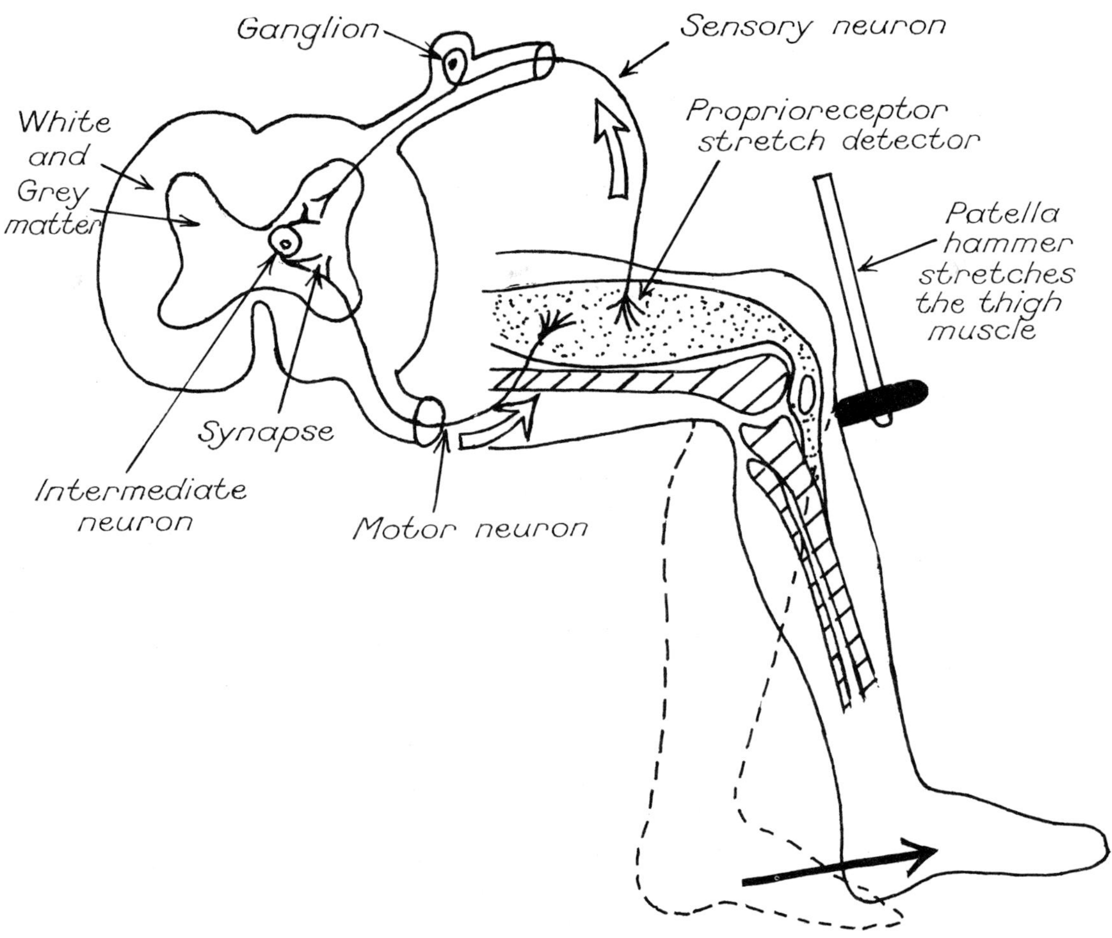

Fig. 10.10 Knee jerk reflex arc

RESPONSES SEEN IN PLANTS AND ANIMALS

Quick automatic responses shown by animals to an external stimulus are called *reflex actions*; the responses shown by plants are called *tropisms*.

Reflex action is the simplest pattern of behaviour seen in animals, involving the receptor or sensory nerve and the effector or motor nerve, both of which are linked by a synapse in the central nerve cord.

This reflex arc is shown in Fig. 10.10. Tapping the area below the kneecap causes the *proprioreceptors* of the thigh muscle and ligaments to stretch and send an impulse along the sensory receptor nerve to the nerve cell body in the grey matter of the nerve cord. The nerve impulse crosses the synapse and continues along the effector or motor nerve which secretes acetyl choline on the nerve plate of the thigh muscle fibres causing contraction of the muscle and the jerking of the lower limb upwards.

The brain has motor and sensory *areas*; in addition it has *memory* stores. It may become involved in a reflex action—called a cerebral reflex action. The three main regions of the brain are:

(a) *Cerebrum:* concerned with intelligence, behaviour and memory. It interprets sound, sight, smell and touch which are located in definite areas over the cerebrum.

(b) *Cerebellum:* the centre for co-ordinating balance, movement, and posture, and orienting the body in space.

(c) *Medulla oblongata:* the centre controlling breathing and the heart beat.

TROPISMS

These are the directional growth movements in a plant in response to an external directional stimulus such as to water (*hydrotropism*) and gravity (*geotropism*). Experiments can be performed showing the bending movement of a plant towards or away from these stimuli shown in Fig. 10.11.

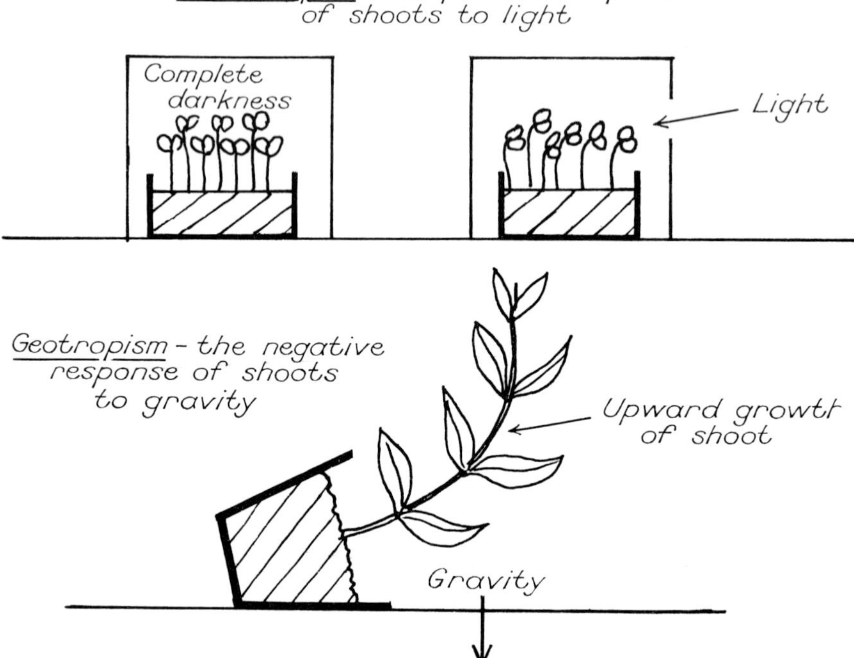

Fig. 10.11 Plant tropisms

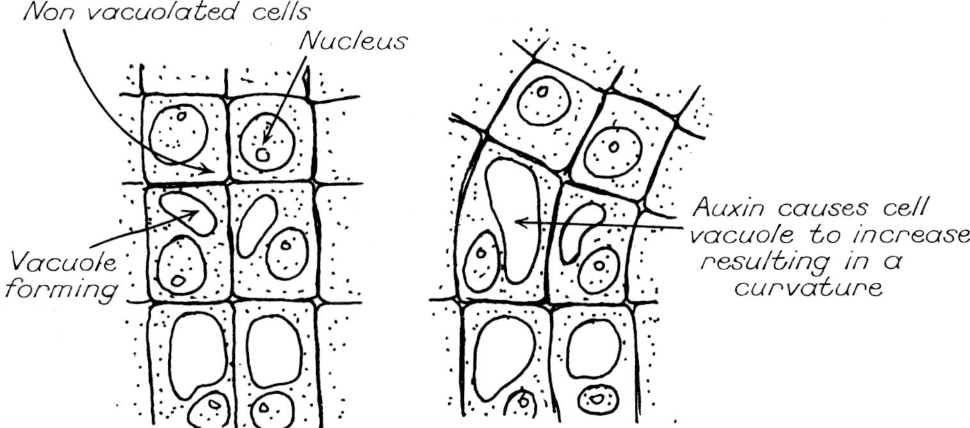

Fig. 10.12 Vacuolation and the resulting curvature it produces

Plants, being without a nervous system, have their tropic movements explained by the plant hormones, *auxins*. These hormones are produced in the cells in the tips of roots or stems, or at the *apices*. The root or stem *apex* or tip receives the stimulus and produces auxin which is passed back to the *growing region*. There the auxin causes cells to increase in size by *vacuolation* the process in which vacuole develops in otherwise dense non-vacuolated cells. As the auxin effects only one side it results in a bending of the stem or root towards or away from the stimulus.

11 *Man and his Environment*

Plants and animals affect their external environment and can also affect each other through the *biotic* environment. Forests and woodlands are examples of *habitats* in which plants and animals live together as a *community*. Man is seen to be responsible for destroying these natural communities—for example in the removal of woodlands to build factories and homes.

WORLD POPULATION

A steady increase took place in the world population from the year AD 1 when an estimated population of from 200 to 300 million, grew into a population of 3600 million in 1970. The population is forecasted to be an expected 7000 million in the year AD 2000. The world population is therefore expected to *double* itself between now and the year 2000.

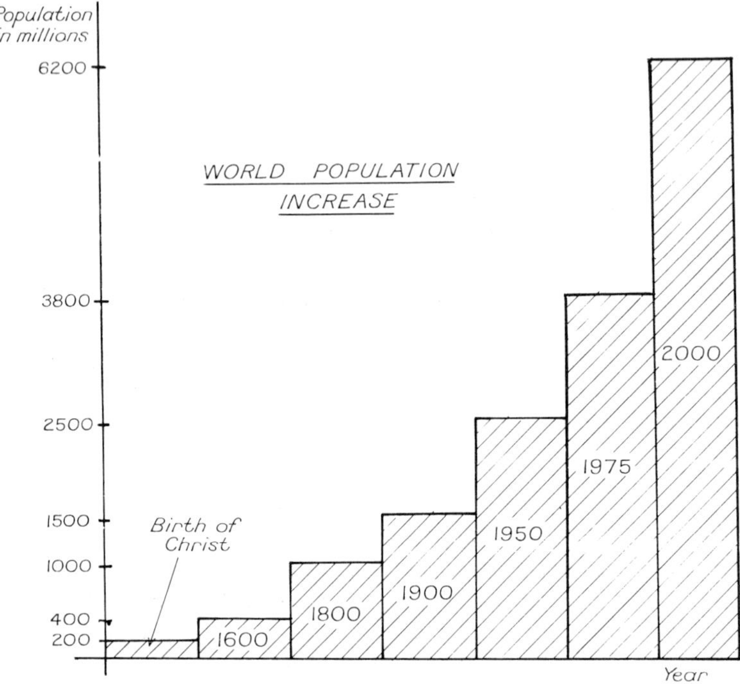

Fig. 11.1 World population increase

Hitherto the increasing world population has been checked by *disease, famine, war* and by various *disasters* such as earthquakes, and major accidents. Many of these *natural checks* in the increasing population have been removed by the progress of medical science in overcoming disease, whereas famine, wars, and disasters still occur.

The *twentyfold* increase in the numbers of man since the year AD 1 has resulted, by man's industrial activities and his over-population, in the *pollution* of his environment. The main ways in which this pollution has taken place is summarised as follows:

(a) Land usage

Forests have been removed for fuel and farmland, upsetting the natural communities of plants and animals. Destruction of the hedgerows is a consequence of the need to grow one type of crop in large fields; this deprives living things of another habitat. The filling in of ponds to level the land for building purposes is yet another loss of a natural habitat.

Towns and cities continually expand to accommodate increasing populations; this leads to the usage of land for factories, roads, railways, airfields, and homes. Approximately 40% of the world's population live in towns and cities.

(b) Pollution of the environment

The manufacturing and farming activities of man lead to the pollution of air, soil, and water.

(1) **Air** is polluted by smoke, dust, and *sulphur dioxide* from burning coal and oil fuels. Smoke and dust prevent beneficial ultraviolet rays reaching the skin of town dwellers to form vitamin D, a deficiency of which can lead to rickets. Sulphur dioxide is harmful to the lungs.

Motor cars and manufacturing industries produce chemical waste products such as *carbon monoxide*, certain cancer-forming *hydrocarbons, rubber* from tyres, *asbestos* from car brakes and *lead* from petrol; many of these are known to be responsible for certain diseases of the lungs and blood.

Pollution by *cigarette* smoke is known to be harmful to the lungs and heart.

Noise in the air from road and air traffic, manufacturing processes, and by loud 'pop' music can lead to psychological disorders and disorders of hearing.

(2) **Water** in seas, rivers and lakes is contaminated by industrial effluent, chemical waste, human waste and sewage.

Poisonous metals *mercury, lead, chromium*, and *cyanide* salts have been discharged as factory effluent into rivers, killing aquatic life and nearby grazing animals, or they have accumulated in the flesh of fish and shellfish to become harmful as food of man.

Plate 30 A comparison showing air pollution: the photographs were taken in Pittsburgh USA, firstly in the 1940s, and secondly after introducing smoke abatement laws in 1970 (World Health Organisation)

Untreated sewage will contaminate shellfish; this form of pollution has been responsible for epidemics of *cholera* and *typhoid.*

Nitrate *fertilisers* are washed out of soils along with harmful microbes from refuse dumps to pollute streams and underground water wells. The increasing use of soapless detergents which have a high phosphate content leads to a situation in which tiny algae multiply rapidly in the water rich in nitrate and phosphate plant nutrients. Rapid multiplication of the algae leads to competition for the dissolved oxygen; ultimately many algae die and their remains pollute the water by this process of *eutrophication.*

Oil pollution at sea has led to destruction of birds, fish, shellfish, and marine plant life.

(3) **Soil.** Intensive methods of agriculture aimed at producing a single crop may involve the use of weedkillers or *herbicides* to destroy the weeds, and *pesticides* and *insecticides* to destroy harmful insects and fungi which would damage the crop. The insecticide DDT has been found to accumulate in the fatty tissues of animals including man to cause harmful effects; for this reason the use of DDT has been banned in the UK. Accidental deaths from weedkiller and pesticide poisoning take place at the present time in increasing numbers.

'Factory farming' of pigs, hens, and calves may involve the disposal of raw animal and silage waste into rivers, and can alternatively be spread over a small area of land causing poisoning of the soil. The overintensive cultivation of the soil results in the soil being eroded by wind, due to the lack of humus content caused by excessive use of chemical fertilisers.

Plate 31 Waste acid being disposed at sea in deep water where it is neutralised, churned and diluted to become harmless to marine life (Bayer Chemicals Ltd)

(c) Pollution by radiation

Atomic radiation is produced by *radioactive* materials which are now finding use as energy sources in nuclear power stations, as alternative energy supplies to natural fossil fuels, coal, gas and oil, which are being used up at a rapid rate and cannot be renewed.

Radioactive materials can produce lethal effects in large doses, whilst repeated exposure to small doses can cause cancer or abnormalities in mammalian embryos including man.

Natural radioactive radiation sources are seen in the rocks and the cosmic radiation from the sky.

Atomic weapons cause pollution by 'atomic fallout' whilst further pollution is possible by careless disposal of radioactive waste from nuclear power stations. Radioactive strontium, Sr90, is produced by atomic bomb fallout and can be passed to man via milk from grazing animals; it accumulates in the bone and attacks the bone marrow causing *leukaemia*.

X-rays, and other radioactive radiations, can 'escape' when used in medicine and industry unless strict precautions are taken. At present, radioactive waste is disposed of in old mine workings, or buried in concrete containers at sea.

(d) Wildlife destruction

Trees have been subject to the greatest destruction by man as is evident by man's concern in the 1973 'plant a tree campaign'. Several flower species are now extinct and many others are now *protected* species.

Hunting activities has caused the extinction of the *dodo* and *auk*, whilst the *golden eagle* is becoming rare in the UK. *Giant pandas* are becoming rare, whilst whales, bison, otters, and seals have been reduced in great numbers. *Overfishing* has caused many countries to extend their territorial waters to conserve fish stocks as food for future generations.

CONSERVATION

In order to retain the purity of the water, air, and soil, to prevent the destruction of wildlife, to protect the health of man and to retain some of the natural sources of fossil fuel, for future generations of man, it is essential to introduce *conservation.*

(a) **Birth control** is needed to reduce the increasing world population, for it is man alone that has caused the pollution in his external environment through over population.

(b) **Land conservation** is needed for food production and is controlled by town and countryside planning authorities.

(c) **Pollution control** is affected by various laws and acts—for noise abatement, clean air, and effluent disposal—implemented by local public health authorities.

(d) **Wildlife conservation** is the concern of various conservancies for nature, game, forestry, and wildlife, whilst international agreements attempt to control fishing and whaling. Various laws prevent the taking of birds' eggs and certain wild flowers.

(e) **Health education** of the individual and the intelligent use of increasing leisure time involves the improvement of the physical and mental health of people living in the overcrowded conditions of towns and cities with its consequent effects on society through tobacco smoking, alcoholism, drug taking, loneliness, aggression and increasing crime.

(f) **Energy** conservation through saving of electricity, coal, smokeless fuel, gas and oil is vital for supplies to continue for future generations.

(g) **The individual responsibility** of every man is to respect the earth and his environment and to leave it a fit place for his descendants.

Plate 32 An instrument to detect and record the presence of air polluting gases in the vicinity of a chemical producing works (Bayer Chemicals Ltd)

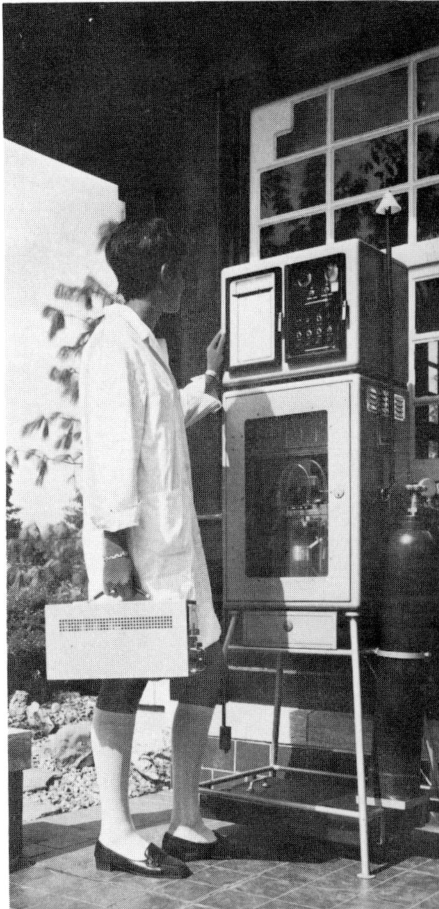

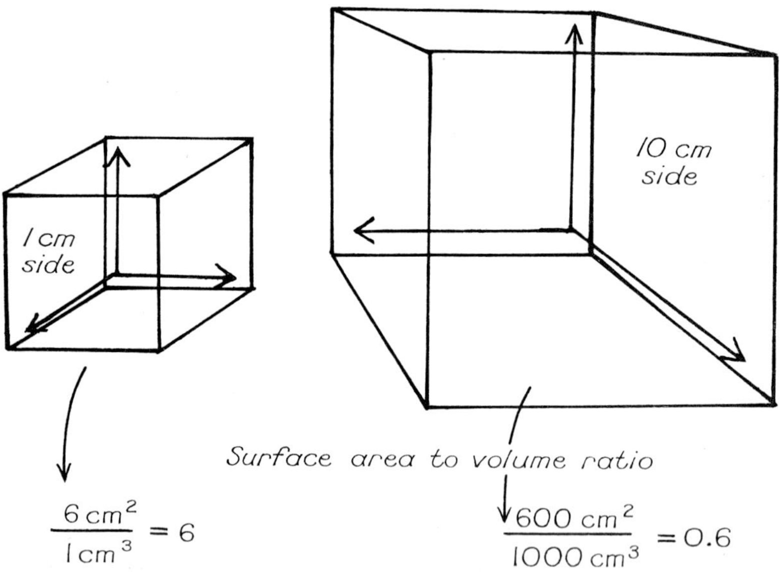

Fig. 11.2 Surface area to volume ratio increase decreases as the volume of an organism increases

GROWTH AND BODY ORGANISATION

Growth of *unicellular* organisms is seen as an increase in *weight*, *volume* and *surface area*; this growth cannot continue indefinitely but is limited to a certain *maximum* size. This limit is imposed by the *reduction* in the ratio between the surface area and body volume as the cell size increases, making less surface area available for exchange of gases and for excretion. When the maximum size is reached, the unicellular organism will undergo *fission* or cell division, to produce daughter cells.

Multicellular organisms commence life as a *single* cell or zygote, which proceeds to divide by mitosis producing a many celled adult. Growth in multicellular organisms is a process of *change* in *weight*, *size*, or *shape*, as a result of nuclear division by *mitosis*.

Growth in multicellular organisms is recorded by:

(a) noting the change in *body weight*.
(b) measuring the increase in *length*, *height* and *girth*.
(c) recording the change in *dry* weight of plants, by taking a disc of plant tissue, drying off the water at 110°C, and noting the change in weight of the dried sample. The *auxanometer* is an instrument used to record plant growth (Fig. 11.3).

Growth is not always seen as an *increase* in weight since many seeds for example rapidly *lose* weight in the early stages of germination when the growth in size is proceeding at its most rapid rate.

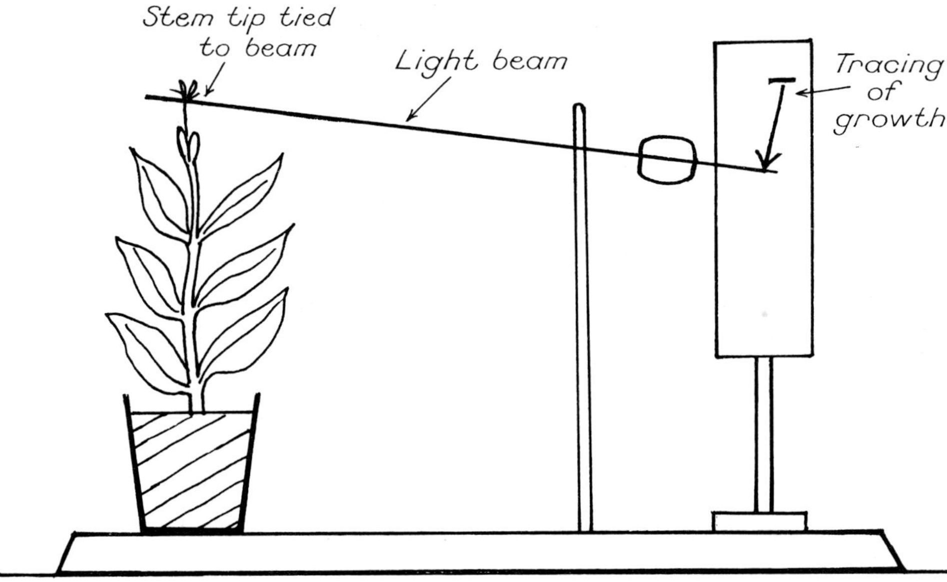

Fig. 11.3 An auxanometer

PHASES OF GROWTH

Growth in multicellular organisms occurs in five phases.

Cell multiplication

In this phase cells divide by ***mitosis***, produces many cells similar to each other or ***undifferentiated.*** In plants this active cell division takes place in regions called the ***meristems***, at the tip of the root and stem. In animals it is not located in any special growing region.

Increase in size

The increase in size of the cells takes place in plants by a process of ***vacuolation*** in the elongating regions of stem and root. A cell vacuole forms in the cytoplasm during this stage (Figs. 11.4 and 10.12, p. 157).

Differentiation

This is a process in which cells become changed to perform special functions in the different ***tissues*** of the plant or animal. In the human embryo differentiation occupies the first three months of its life forming the primary layers of the ***ectoderm, endoderm*** and ***mesoderm.***

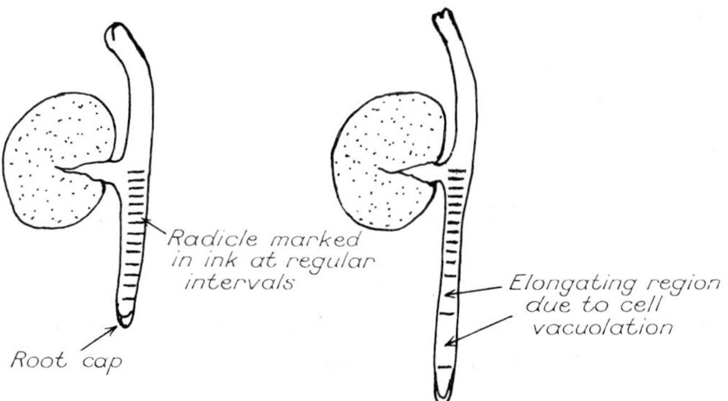

Fig. 11.4 Increasing cell size in the elongating region of a plant radical

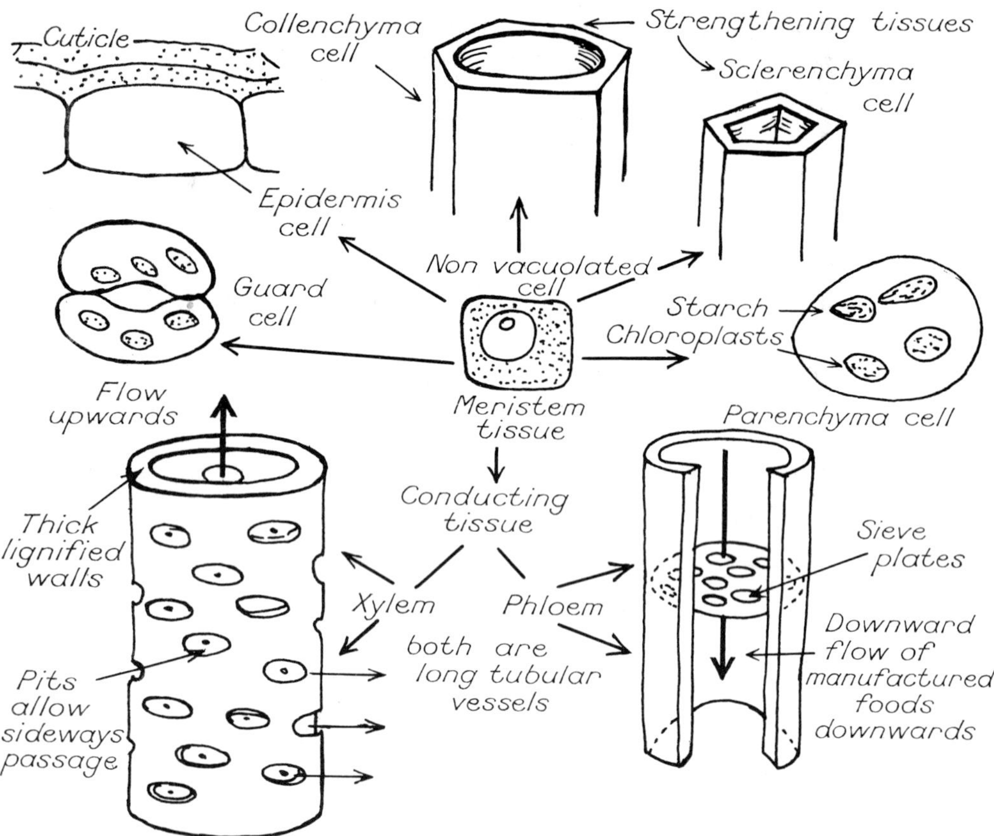

Fig. 11.5 Cell differentiation from meristem tissue in plants

Differentiation in plants. This proceeds as follows—from the meristem cells:

(1) *Parenchyma* is a tissue consisting of mainly thin-walled cells either storing food or containing chloroplasts.

(2) *Strengthening and supporting cells* have specially thickened walls in tissues called the *sclerenchyma*, and *collenchyma.*

(3) *Vascular tissue* of the vascular bundles consists of vessels and *tracheids* found in the *xylem* or the wood, and *sieve tubes* and *companion* cells are found in the *phloem* or bast of the vascular bundles.

(4) *Guard cells* or stoma are important in allowing gases to enter and leave the leaf. (See Fig. 8.15, p. 126, and Plate 29.)

(5) *Epidermis cells,* of stem, root, and leaf, are mainly protective and prevent drying out of the tissue beneath. Plates 33 and 34 (pp. 168–9) show how these tissues are located in the stem and root.

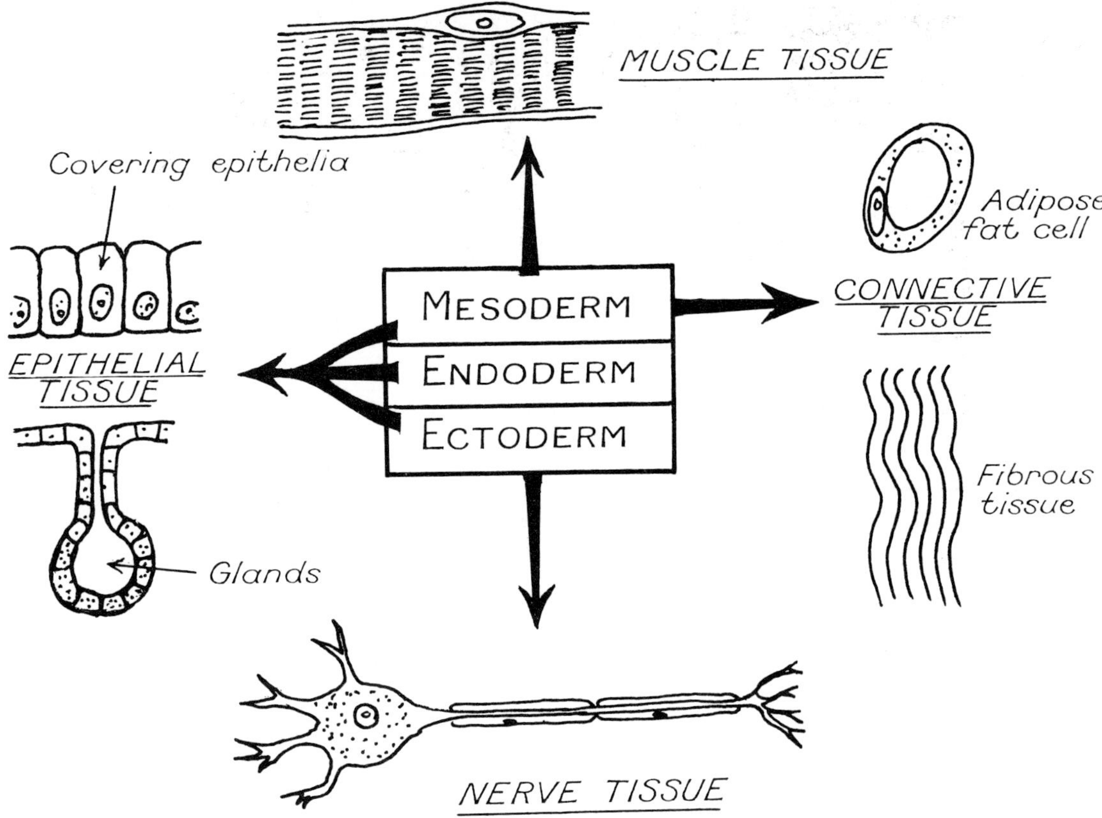

Fig. 11.6 Differentiation of primary germ cell layers into different tissues in animals

Differentiation of tissue in animals. This proceeds as follows—from the three primary germ cell layers, *ectoderm*, *endoderm*, and *mesoderm*:

(1) *Epithelial tissue* covers and lines the body and its glands; it forms mainly from the ectoderm and can also form from endoderm and mesoderm.

(2) *Connective tissue* of the bones, cartilage, ligaments, and tendons serves to support, bind and enswathe the body: it also includes the blood. All of this forms from the mesoderm layer.

(3) *Muscle tissue* of the body limbs, heart and intestine forms from the mesoderm layer.

(4) *Nerves* and nervous tissue of the brain and spinal cord form entirely from the ectoderm layer.

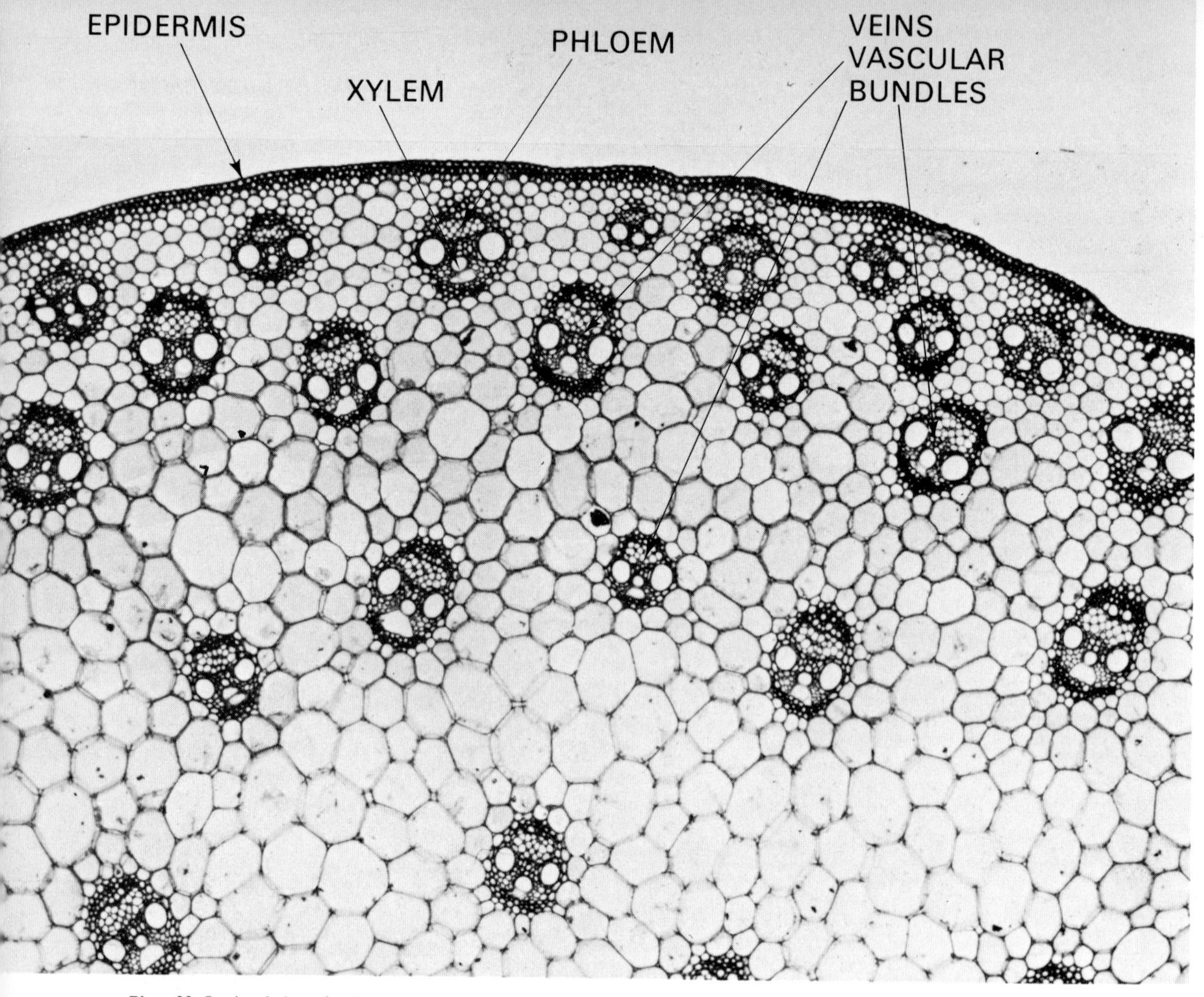

Plate 33 Sectional view of a plant stem seen through the microscope showing the numerous vascular bundles and different cell tissues (Griffin Biological Laboratories)

Plate 34 Sectional view of a plant root showing the main tissues with a central vascular tissue (Griffin Biological Laboratories) (facing page)

Active growth period

This is the period of growth when a considerable increase in length or height occurs, in the young plant or animal. This occurs in plants from the time the plant has formed its first leaf until the formation of fruits; in human beings it extends from the third month of embryonic development until the age of 20 years. The growth period varies in different animals: for example it occupies 10 years in monkeys and apes.

In animals it is essential to have growth limits as it is believed that the extinction of reptiles was due to their unwieldy size caused by excessive growth. Trees do not appear to have growth limits or curbs; this is seen in the great age attained by certain trees, e.g. Yews, and in the giant size of Canadian Redwood trees.

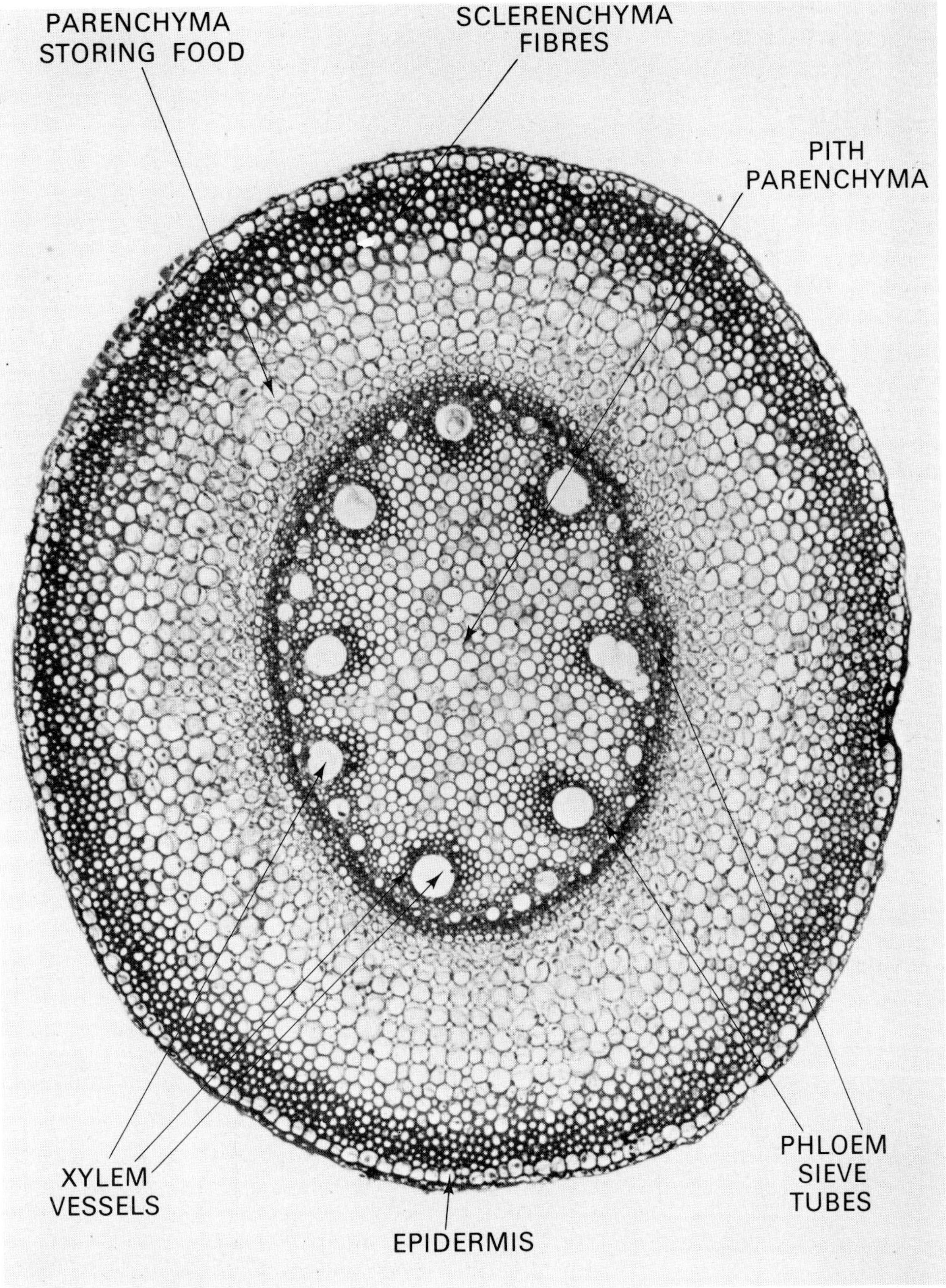
PARENCHYMA
STORING FOOD
SCLERENCHYMA
FIBRES
PITH
PARENCHYMA
XYLEM
VESSELS
EPIDERMIS
PHLOEM
SIEVE
TUBES

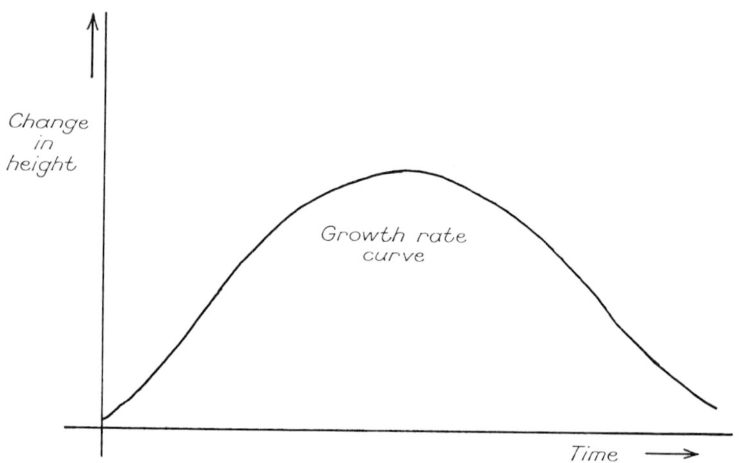

Fig. 11.7 General growth rate curve for living organisms

Senescent or static period

Plants and animals enter a period of very slow or limited growth proceeding old age and death. Frequently there is a period of weight loss. Man can show a reduction in height due to shortening of the vertebral column.

GROWTH CURVES

The growth rate of living organisms can be displayed in *graphs* which show a curve that generally follows a pattern in which the curve rises to a maximum and declines towards the end of the organism's life, as shown in Fig. 11.7.

Growth in plants

Growth in plants is followed by changes in the *dry weight*; in the early stages seedlings show a rapid loss in weight due to depletion of food reserves in the seed. The active growth period results in a steady increase in weight due to cell division and differentiation with photosynthesis food manufacture taking place. After flowering and fruiting, the plant shows a *loss* in dry weight whilst foodstores are used up.

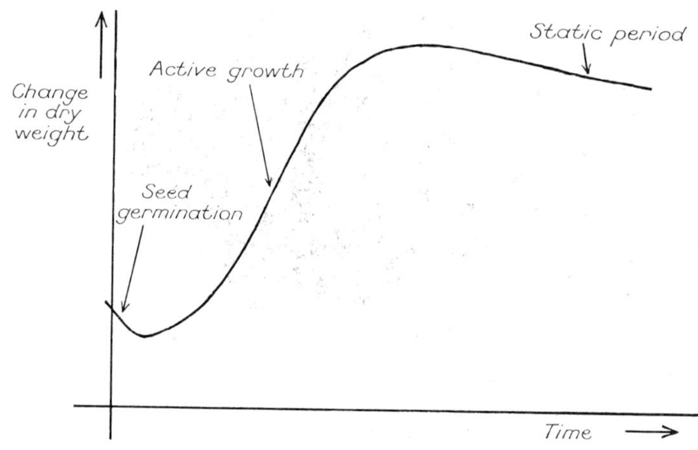

Fig. 11.8 Growth curve for flowering plants

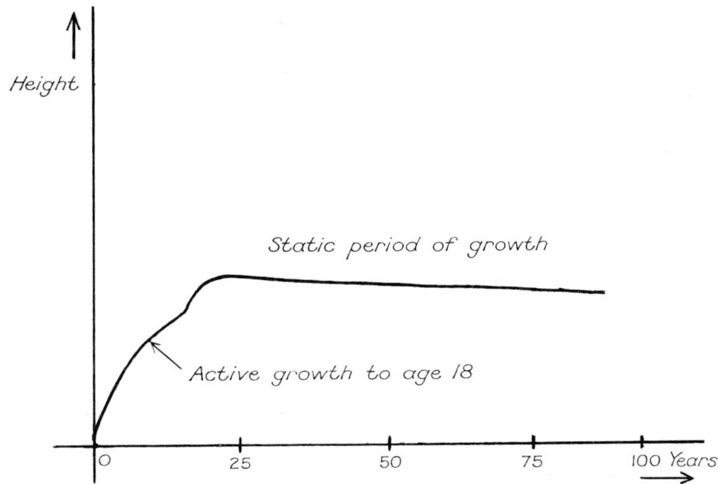

Fig. 11.9 Growth as a change in height in man

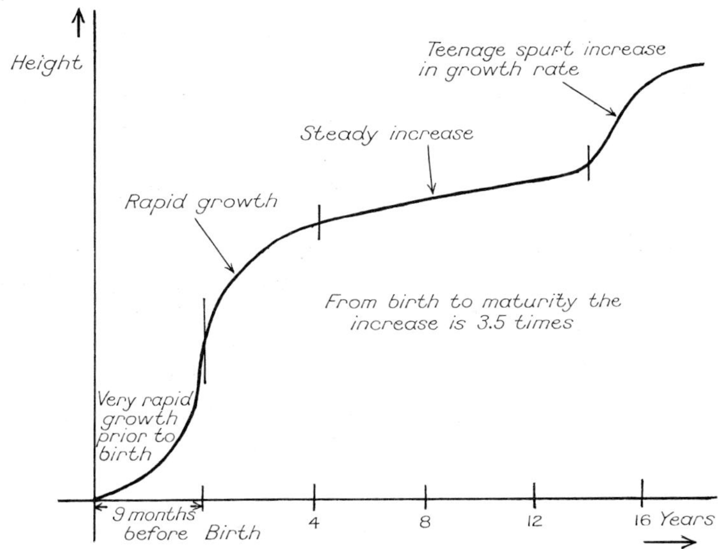

Fig. 11.10 Growth of man from before birth to age eighteen

Growth in man

Growth in man is recorded by changes in height, weight, and surface area of skin, all of which show an increase up to the age of 20 years.

The graph in the Fig. 11.9 shows the change in *height* during the lifetime of man; the growth is most rapid in the human embryo in the 6 months prior to birth and up to the age of 2 or 3, after which growth takes place at a fairly steady rate up to the age of 12 years. An increase in the growth rate is seen to occur at *puberty* from about 12 to 17 years, after which the rate of growth slows down until the age of 20 to 21 when it enters the static period (Fig. 11.10).

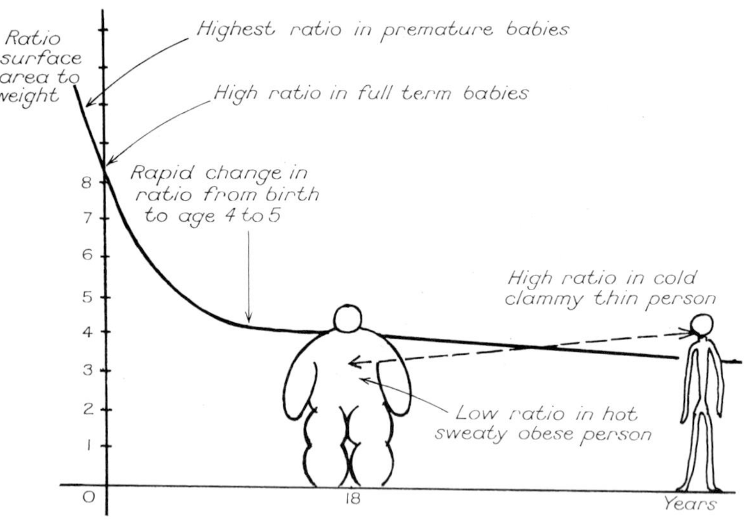

Fig. 11.11 Change in the ratio of surface area to weight ratio from birth to adulthood

Changes in the *ratio* of skin surface area to body weight or volume is shown in Fig. 11.11 when the *highest* ratio is seen in the small babies, compared to the adult man. Babies with a high skin surface area to weight ratio will lose heat rapidly if unclothed. A thin underweight man will feel cold and chilly, in contrast to a short fat man who will feel hot and sweaty. Adults of normal weight and height will have a ratio of surface area of skin to body weight that should maintain the body at a comfortable temperature in normal surroundings.

GROWTH REGIONS

Plants show an increase in *girth* which is seen in the *annual rings* of growth inside the stem of trees; this is produced by new growth of *vascular tissue* occurring each year in the spring from a meristem called the cambium. A count of the annual rings will give an estimate of the tree's age. Similarly growth *rings* are seen in fish scales, *otoliths* in the ear, and in mollusc shells (Fig. 11.12).

Animal body growth occurs at different rates in different parts of the body. In man the head of the newborn child and the growing child develops at a greater rate than the rest of the body; this is due to the greater development of the brain in comparison with other organs.

Mammal bones are limited by the growth of the *epiphysis* cartilage, which becomes hardened or ossified into bone and therefore halts bone growth after 18 or 19 years in man (Fig. 11.13).

GROWTH AND ENVIRONMENT EFFECT

Plant growth is mainly affected by the external environment in the effects of light, temperature, water and oxygen. In total darkness the growth of plants is rapid; this produces tall, thin, weak plants devoid of chlorophyll.

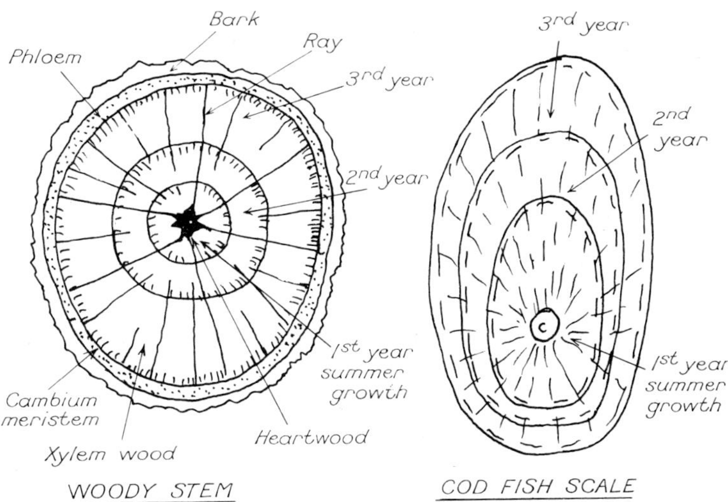

Fig. 11.12 Annual growth rings seen in plants and animals

Fig. 11.13 Epiphysis cartilage of young animal bone which allows growth in length and girth of the bone

The internal environment affects the growth rate of living organisms mainly through the effects of changes in *food supply*, *hormones*, and *genetic* influence of the cell nucleus.

Since growth is a result of the energy released by food together with the building up of the body from certain proteins and minerals, it will be evident that food and nutrients have a major effect on plant and animal growth. Plants which suffer deficiency of nitrogen, phosphorus, and potassium in the soil nutrients show poor growth. Animals with deficiencies in their diets have their growth affected. In man, rickets is due to vitamin D and calcium deficiency and Kwashiorkor is due to protein deficiency.

Hormones can affect growth in plants (as seen in the effect of *auxins*) whilst in animals disorders of the pituitary gland can cause gigantism or dwarfism. Defects of the thyroid gland can produce *cretins*, and sex hormones *androgen* and *oestrogen* cause the growth spurt seen in puberty. *Cortisone* can retard the growth rate.

The cell nucleus contains *genes*; these have a profound effect in producing inherited tallness or shortness in plants and animals. This subject is discussed in the next chapter.

12 Genes and Genetics

The classification of animals was outlined in Chapter 2, the following illustrates the more detailed classification of man, the domestic cat and dog into a *genus* and *species*:

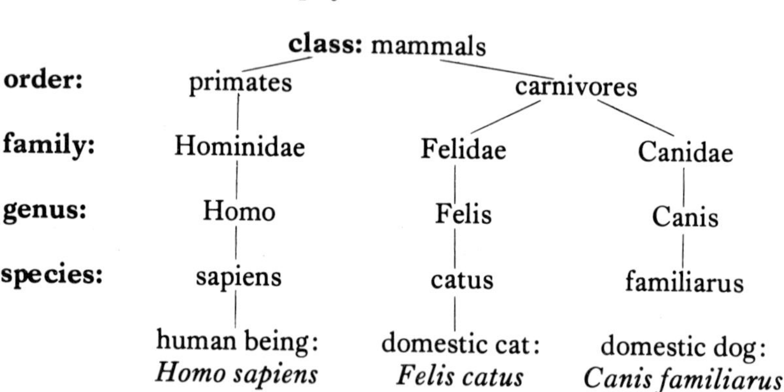

GENUS AND SPECIES

Animals with a spine or backbone are members of the phylum *chordates*. As all the animals shown above suckle their young with milk from mammary glands they are included in the same class as *mammals*. Cats, dogs, and human beings show differences which place them in different *families*, man being a member of the *Hominidae* as described in Chapter 5. Each family is subdivided into *genera* and each is allocated a *species*. The genus name is given a capital letter whilst the whole name is underlined or printed in italics.

Members of the *same species* can interbreed with each other, but members of *different* species cannot normally interbreed, if they do the offspring will be *sterile*: for example a horse and donkey will produce a *mule*, whilst a lion and tiger will produce a *tigon*. Both of these offspring being sterile, cannot reproduce their kind. The test of interbreeding, used to determine the purity of a species, is the ability to produce a *fertile* offspring.

Similar classification is used for plants. Peas and beans are both members of the same family—the *legumes*; peas are of the genus *Lathyrus*, and beans are of the genus *Vicia*, neither of which can

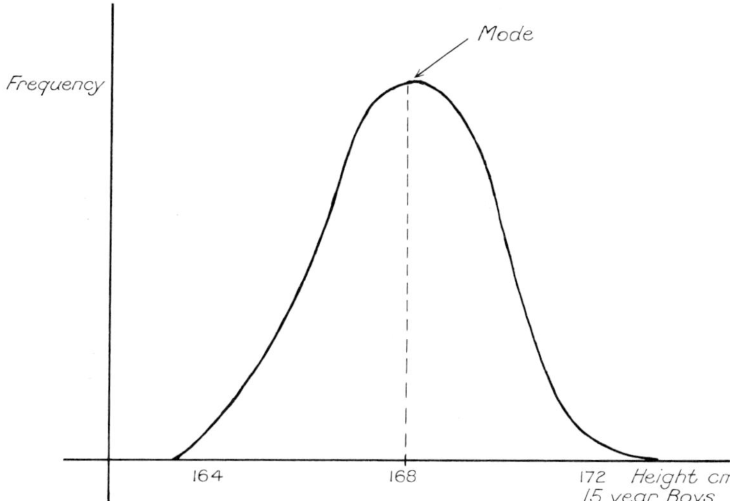

Fig. 12.1 Continuous variation in a height distribution curve for 15-year-old boys

interbreed. Crossbreeding is possible amongst peas to produce such *varieties* as the edible garden pea and the pleasant smelling sweet pea.

VARIATIONS

Members of a particular *race* of man can show a number of differences between the individual members of the race, these differences can also be recognised amongst individual members of a domestic family, apart from the almost indistinguishable appearance of *identical twins*. Differences between individual members of the same species of plants and animals are called *variations*.

(a) **Continuous variations** are *gradual* or graded differences between two extremes within the same species; this is seen in *population statistics* as a difference in such things as height, weight, leg and chest measurements. These variations seen in man are of a gradual nature: for example the heights of 15-year-old boys can range between 150 and 180 cm within a population. If a graph is drawn a *distribution curve* resembling a bell shape will be seen with most of the 15-year-old boys being around the *mode* of 168 cm in height (see Fig. 12.1).

Several reasons can be given for this variation in height. The individuals may be differently fed or differently housed, which are *external* environmental influences. Otherwise they could be the children of tall or short parents from different races, their height being controlled by the *internal* environmental effect of the *genes* in the cell nucleus.

(b) **Discontinuous variations** are seen very clearly and distinctively as differences in the species of plant or animal; for example the inability of certain people to see certain colours, taste certain

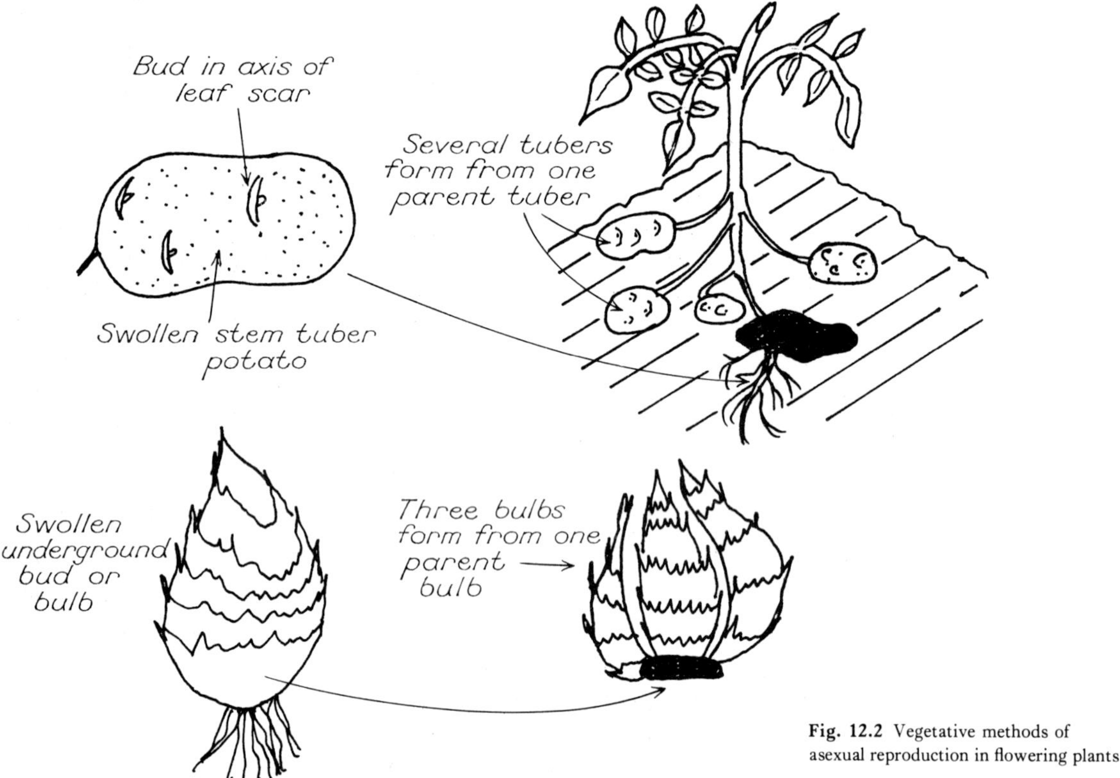

Fig. 12.2 Vegetative methods of asexual reproduction in flowering plants

chemicals, e.g. phenyl thiourea, or the inability to roll the tongue into a certain shape, or in the absence of lobes to the ear. These discontinuous variations are *inherited* or inborn. This inheritance is decided by the *genes* of the nucleus and cannot be influenced by the external environment. Several discontinuous variations are seen as serious hereditary diseases in children as *phenylketouria*, and *maple syrup* disease, whilst the inheritance of the human blood groups allows *one* choice of either group A, B, AB, or O.

REPRODUCTION

Reproduction is by two main methods; *asexual* or vegetative reproduction in which *one* individual is involved, and *sexual* reproduction in which *two* individuals are involved, one male and one female.

Asexual reproduction

This method of reproduction is seen in animals as *binary fission* in the *Amoeba*, or as budding in *Hydra*. Both are processes of division through nuclear division by *mitosis*. The offspring will be identical to the parent cell or animal in most respects and will show little variation apart from environmental factors affecting size, etc.

Asexual reproduction is seen in plants which employ vegetative methods using bulbs, corms, rhizomes and runners.

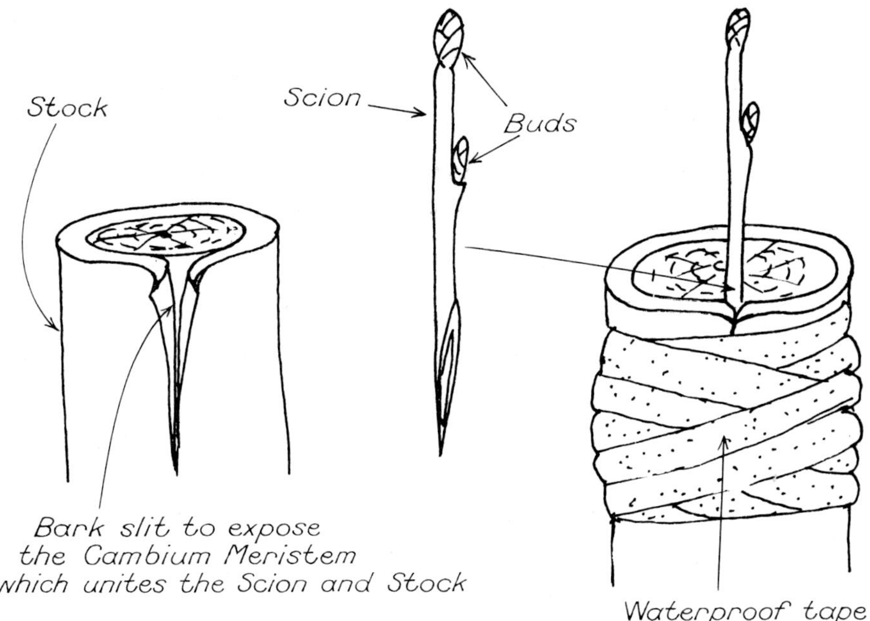

Fig. 12.3 Artificial propagation by grafting

Artificial methods of propagation using cuttings, grafting or budding will produce offspring nearly identical to the parent scion plants. Vegetative reproduction preserves the main features of the parent scion plant and both beneficial and harmful characteristics will be passed on directly to the offspring.

Clones are plants and animals produced from a common ancestor by mitotic division; such clones are seen in different variety of apple. Cox Orange Pippin apples for instance have all been raised from one tree and propagated by scion cuttings and grafting.

Identical twins are considered as clones since they have developed from *one* zygote cell, after it has undergone the first mitotic division; each cell develops into an embryo that will be identical to the other having been formed from the same cell and nucleus material.

Fraternal twins are produced as *dizygotic* twins by the fertilisation of *two* separate egg cells, forming two zygotes which develop into embryos showing some variation from each other since they are formed from different cell and nucleus material.

Sexual reproduction

Male and female *gametes* from different parents fuse together and form zygotes by a process of *cross fertilisation*; the gametes come from different plants or animals of the same species. *Self fertilisation* is a process in which gametes from the same organism, e.g. a flower, fuse together to form a zygote. Self fertilisation produces *pure lines* of organisms, compared to cross fertilisation which allows variations to appear in a mixed or *hybrid* species.

Sexual reproduction produces an assortment of *characters* in the offspring which may be seen as continuous or discontinuous *variations*, in contrast with the mainly uniform appearance of all offspring produced by asexual vegetative reproduction.

HEREDITY

Parent plants and animals pass on certain features to their offspring; these inherited features are called *characters.* As each parent provides the inheritable characters they may be either *similar* or *contrasting.* Inherited characters are seen as flower *colour*, or eye and hair colour.

Johann Mendel (1822–1884), an Austrian botanist and abbot of the Augustinian monastery at Brúnn, investigated the inheritance of characters in plants; the results of his investigations, mainly with pea plants, founded the study of *heredity* and *genetics* also called *Mendelism.*

DOMINANT AND RECESSIVE CHARACTERS

The fundamental results of Mendel's experiments in plant breeding are summarised as follows:

(a) **Pure bred** pea plants are those plants which have continually produced seeds by *self fertilisation* to provide a pure line of peas. Self fertilisation is brought about by completely covering individual flowers with bags to prevent pollen from other flowers reaching them.

(b) **Characters,** such as *tallness* or *dwarfness* seen in pea plants, are selected and pure bred lines of these are obtained by continual self fertilisation of the flowers, until only pure tall or dwarf plants are produced consistently from the pure bred seeds.

(c) **Cross fertilisation** of pure bred strains of pea plants involves the transfer of pollen from tall pea plants to the flower stigmas of dwarf pea plants, and vice versa. This requires a special camel hair brush and great care; after pollination the fertilised flowers are completely covered with bags to prevent fertilisation by stray pollen.

(d) **Determination of dominant character** is made by carefully collecting all the seeds from the pea pods, sowing them and allowing them to grow fully into adult plants. If *all* the plants turn out to be tall this will indicate that tallness is the *dominant* character, in contrast to dwarfness which is the *recessive* character. The plants which have grown from the seeds are called the *first filial* generation or F1 generation.

(e) **Self fertilisation** of the first filial generation is carried out on the flowers of the F1 plants, taking care that no cross fertilisation occurs. After the pea pods form, the seeds are collected and are sown to germinate into fully grown plants of the *second filial* generation or F2 generation. An examination of the fully grown plants show some to be tall and some are dwarf, in a ratio of *three* tall plants to *one* dwarf plant.

SUMMARY

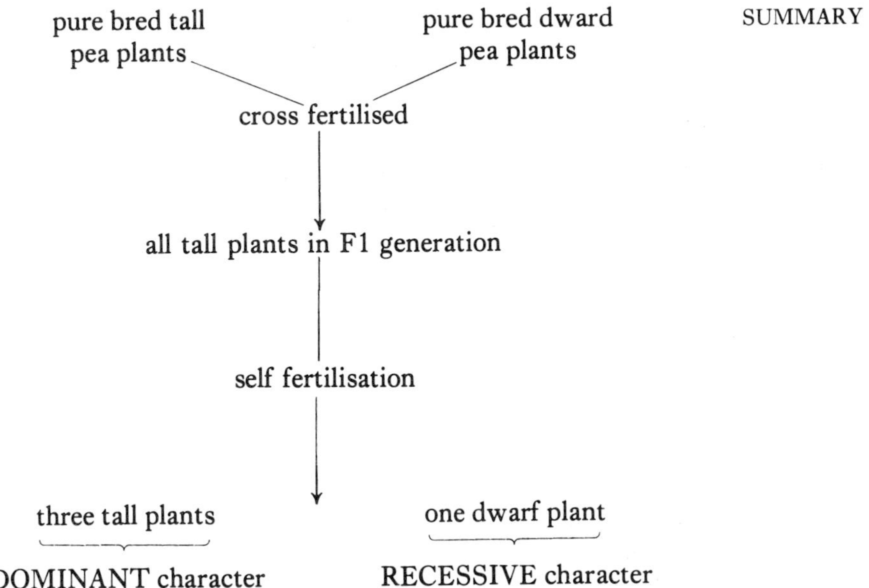

(1) **Dominant** inherited characters will show in the first filial generation F1 after crossing *pure* bred lines.

(2) **Recessive** characters will be seen in the F2 generation after self fertilisation.

CHROMOSOMES AND GENES IN INHERITANCE

Cell nuclei are composed of *chromosome threads*. Each species of plant or animal has a definite number of chromosomes in its nucleus; this is called the chromosome number. Man has 46 chromosomes. Mitosis of *body* cell nuclei duplicates the chromosomes in a regular manner without any changes in their structure. Thus every daughter cell has the same number of chromosomes as the parent cell.

Reproductive cells of plants and animals have a nucleus which divides by *meiosis*; this process, summarised in Chapter 1, involves the chromosome *pairs* separating into *single* chromosomes and is followed by a process of *crossing over* in which a *mixing* or *assortment* of the chromosome material takes place. This stage is important in order to introduce the new *variations* seen in the offspring.

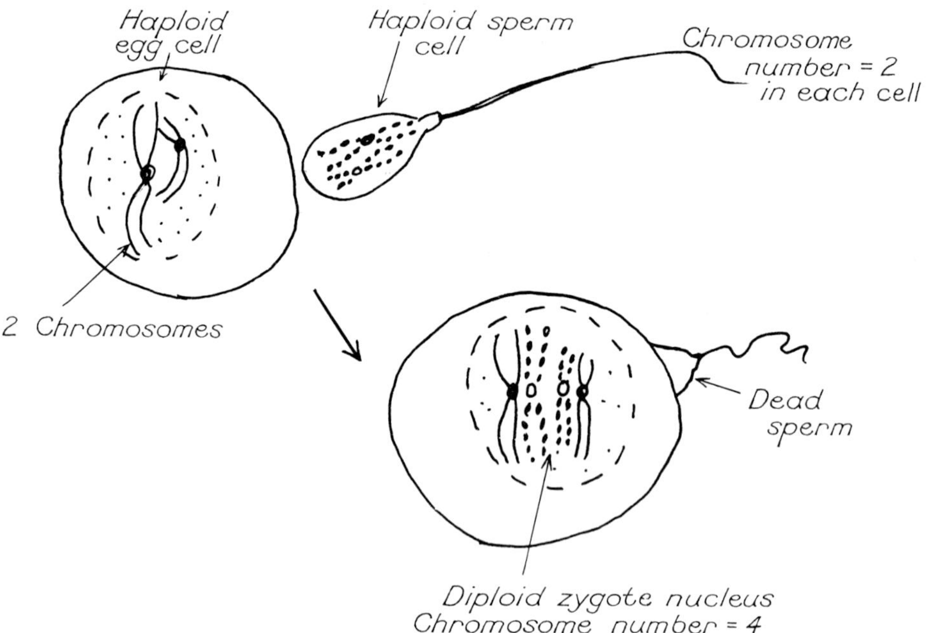

Fig. 12.4 Fertilisation producing a diploid zygote

Fertilisation brings together *single* chromosomes of the gametes—bearing in mind that the chromosome number has been *halved* during gamete formation in meiosis. These *haploid* gametes from *zygotes* with the normal chromosome number of the species; such zygotes are called *diploid. Genes* are found along the length of a chromosome and can be compared in a simple way to lines appearing on a tape measure. The purpose of the gene is summarised as follows:

(a) Genes are *complex proteins* of DNA (deoxyribose nucleic acid) which control the activity of body cell *enzymes*.
(b) They are always *paired* opposite other genes in a pair of chromosomes.
(c) The paired genes may be *similar* or *dissimilar*.
(d) Genes are either *dominant* or *recessive*.
(e) Pairs of similar genes are called *homozygous* whilst dissimilar genes are called *heterozygous*.

CHEQUER BOARD GAME OF INHERITANCE

A chequer board of four squares is a useful means of explaining character inheritance.

Each pure bred tall plant produces male pollen and female egg cells each having a single chromosome gene for tallness, this is indicated by a capital letter **T**, whilst the pure bred dwarf pollen and egg cells have a single chromosome gene for dwarfness indicated by a small letter **t**. The symbols **T** and **t** are arranged on the chequer board and the different combination of genes in the chromosome *pairs* shown as **Tt, tT**, in the squares as follows:

Sex cells of *pure* dwarf plant		Sex cells of *pure* tall plant	
		T	**T**
	t	**tT**	**tT**
	t	**tT**	**tT**

Prediction. Reading the chequer board show that the products of crossing tall plants with dwarf plants will be *hybrids* of the gene makeup **tT**. Such a pair of contrasting characters will show only the *dominant* character, tallness, which overshadows the recessive character for dwarfness.

The chequer board is used to explain the outcome of self fertilisation of the F2 generation as follows:

Sex cells of F1 tall plants		Sex cells of F1 tall plants	
		T	**t**
	T	**TT** HOMOZYGOUS tall plant	**Tt** HETEROZYGOUS tall plant
	t	**tT** HETEROZYGOUS tall plant	**tt** HOMOZYGOUS dwarf plant

Prediction. The chequer board shows the product of self fertilisation to produce the F2 generation. In this there will be *three tall* to every *one* dwarf plant. *Homozygous* plants will have pairs of similar genes in **TT** and **tt**, i.e. a pure tall plant and a pure dwarf plant. The other tall plants will be impure or *heterozygous*, with contrasting genes, **Tt** or **tT**.

PURE BREEDS

The three tall plants **TT, tT, Tt**, *outwardly* show by their physical appearance that they are tall, and are called *phenotypes*; *inwardly* or genetically there are only two *genotypes*, viz. **TT** and **tt** which are also homozygous and of true or pure breeding.

DOMINANT AND RECESSIVE CHARACTERS IN MAN

Several disorders and conditions are inherited due to a single dominant or recessive gene from one of the parents. Some of the inherited disorders seen in man are as follows:

Conditions in which one gene is needed as a dominant character to display the condition

(a) White forelock of hair amongst normal coloured hair
(b) Missing ear lobes
(c) Skin freckles
(d) Extra fingers or toes
(e) Missing fingers or toes
(f) Brown eyes
(g) Ability to taste phenyl thiourea
(h) Resistance to tuberculosis
(i) Huntington's chorea
(j) Fetal rickets

Conditions in which one recessive gene is needed to display the condition

(a) Blue eyes
(b) Red hair colour
(c) Albinism—white hair and skin with pink eyes
(d) Alkaptonuria—a condition in which the urine darkens to dark colour on long exposure to air
(e) Phenylketonuria—inability to digest certain amino acids with formation of phenyl ketones in urine
(f) Certain anaemias and cystic fibrosis disease

SEX AND BODY CELL CHROMOSOMES

Of the total number of chromosomes in a *body* cell of man 22 pairs are called *autosomes*, whilst *one* pair are called the *sex chromosomes*; these are responsible for the *sex* characters of males and females. The sex chromosomes can be seen in photographs or *karyograms* of the nuclear material as seen through an electron microscope (see Chapter 1, p. 11). Karyograms of the female human nucleus show it to have two large chromosomes of a similar size called the X chromosomes (see Plate 35, p. 185), whilst karyograms of the male human nucleus show one large or **X** chromosome and a smaller **Y** chromosome in the pair.

The female cell nucleus in man has a pair of *similar* chromosomes **XX** and is *homozygous*, the male cell nucleus has a pair of different chromosomes **XY** and is *heterozygous*, both of these sex chromosomes influence the development of sex in man through exerting an influence in the production of the hormones.

MALE AND FEMALE

The ratio of females to males in the human population is roughly one female to every male. This distribution can be explained by the chequer board diagram.

Male sex cell chromosomes

		X	Y
Female sex cell chromosomes	X	XX HOMOZYGOUS female	XY HETEROZYGOUS male
	X	XX HOMOZYGOUS female	XY HETEROZYGOUS male

Prediction. The human population will consist of 50% males and 50% females.

CHROMOSOME MISTAKES

Breakages or other damage to the delicate and minute chromosomes can occur and will result in chromosome mistakes to form abnormal or defective chromosomes. It is believed that one in every hundred human embryos have an abnormal chromosome structure. Many abnormal embryos do not survive being lost by *spontaneous abortion.* The embryos which do survive may display the following conditions through chromosome abnormality.

(a) **Down's syndrome,** also known as mongolism, is due to the presence of either an *extra chromosome* making the total number 47 instead of 46, or one extra chromosome in addition to the 22 pairs of autosomes.

(b) **Sex chromosome abnormalities** are concerned with defects arising in the **X** and **Y** chromosome of sex.

Abnormal females may have only one **X** chromosome (making a total of 45 chromosomes) instead of the normal **XX** pair. Such females are infertile and of small stature.

An extra **X** chromosome may be seen in females having a triple **X**, or **XXX** sex chromosomes (making a total of 47); the females are normal in most respects but may show abnormal intelligence.

Abnormal males, instead of having the normal **XY** pair, may either have an extra **X** chromosome in an **XXY** combination (these men are sexually infertile) or possess an extra **Y** chromosome, producing the **XYY** combination (these males are frequently aggressive, sexually normal and of tall stature).

(c) **Polyploid zygotes.** Gamete cells are *haploid* having half the chromosome number; human gametes have 23 chromosomes which produce a zygote with 46 chromosomes, which are *diploid*. Human embryos are occasionally formed with one and a half to two times the normal diploid chromosome number, that is with 69 or 92 chromosomes instead of 46; these embryos are called *triploid* and *tetraploid* and seldom survive since they are spontaneously aborted or miscarried at an early stage in their development.

Many cultivated vegetables, fruits and flowers are *polyploid*, having more than one set of chromosomes; for example there are large showy double wallflowers which are diploid. Similarly certain varieties of tomatoes, raspberries and wheats with large fruits are polyploid.

SEX LINKED INHERITANCE

The **X** and **Y** sex chromosomes carry *genes* like all other body chromosomes; these genes can affect body structure and function and may exert a harmful effect. *Sex linkage* is the inheritance of a disorder through a certain gene influence found on the **X** chromosome.

Haemophilia is an example of a sex linked inheritance. The gene responsible is found at the end piece of the **X** chromosome and can be indicated by * in **X***.

The pairing of a **X*** chromosome with a **Y** chromosome produces a male **X*Y** who will show the symptoms of the disease; this is mainly due to the influence exerted by the haemophilia gene * which is unpaired with any gene in the **Y** chromosome.

The pairing of an **X*** chromosome with another **X** chromosome produces a female **X*X** who will be a 'carrier' but since the haemophilia gene * is paired with another gene its influence is not shown and the female carrier **X*X** shows no symptoms of the disease; however the carrier is able to transmit the disease to her children.

Females can become haemophiliatic and show the disease if an **X*** chromosome pairs with another **X*** chromosome to form a female **X*X*** with a double dose of the gene; such female sufferers are rare compared to the incidence of male sufferers.

Other sex linked diseases include colour blindness, and Duchene's muscular dystrophy disease.

Amniocentesis is the technique of removing fluid from the amnion sac surrounding the embryo in the uterus. The fluid cell nuclei may be examined through a microscope. A dark staining body will be seen under the nuclear membrane in the fluid surrounding normal female babies: this dark staining body is called the *Barr body*. A normal male baby will not show a Barr body in its fluid cell nuclei. Apart from determining the sex of the yet unborn baby, the biochemical examination of the chromosomes may show that certain genetic abnormalities are present (see Fig. 1.1, p. 3).

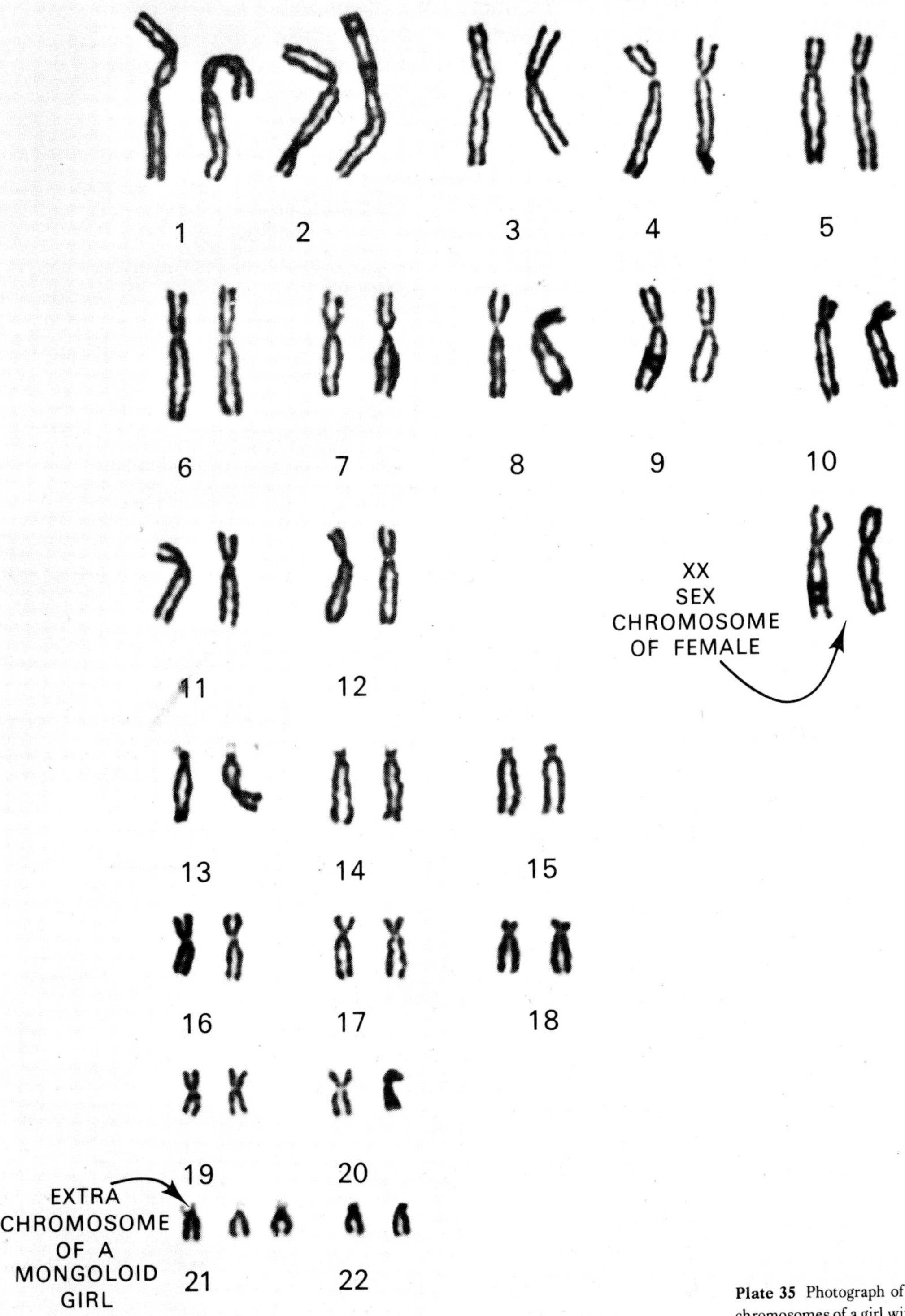

Plate 35 Photograph of the displayed chromosomes of a girl with Down's syndrome (Sandoz Ltd)

MUTATIONS

Permanent and unexpected changes may be seen in the offspring, such as the formation of 'double' flowers in plants or extra fingers and toes in animals. These changes are called *mutations* and are also examples of discontinuous variations.

Mutations are the result of changes within the chromosome structure or in their number, or in the structure of a particular *gene*. Genes and chromosomes are complex chemical compounds and as chemical compounds can undergo chemical changes which in the case of mutations are unable to be reversed.

During meiosis in the nuclear division of reproductive cells, the chromosomes undergo considerable rearrangement; this happens in the *crossing over* stage. At this time there is a mixing of chromosome material and genes may be damaged accidentally.

Changes in the genes can be induced by *external* means through the effect of chemical compounds, drugs such as 'Thalidomide', and mustard gas and colchicine. Radiation such as X-rays, ultraviolet rays, gamma rays and the natural radiation from rocks and soils are also known to cause mutations in the chromosome material of living organisms.

INHERITANCE AND THE ENVIRONMENT

The internal environment of living cells is controlled by the genes on the chromosomes; these determine for example the tallness of a person, or the colour of the eyes and hair. Similarly harmful genes may cause disorders such as *haemophilia*, whilst *lethal* genes will kill the organism which bears the genes in its cells.

The *external environment* may influence the development of variety within a population of organisms but this does not play such an important part as that exerted by the genes which are the main materials of inheritance. Difference may be seen in identical twins reared apart after birth in differing home environments but such differences are only temporary and can be reversed.

Index